Springer Natural Hazards

The Springer Natural Hazards series seeks to publish a broad portfolio of scientific books, aiming at researchers, students, and everyone interested in Natural Hazard research. The series includes peer-reviewed monographs, edited volumes, textbooks, and conference proceedings. It covers all categories of hazards such as atmospheric/climatological/oceanographic hazards, storms, tsunamis, floods, avalanches, landslides, erosion, earthquakes, volcanoes, and welcomes book proposals on topics like risk assessment, risk management, and mitigation of hazards, and related subjects.

More information about this series at http://www.springer.com/series/10179

R. Jayangondaperumal · V. C. Thakur
V. Joevivek · Priyanka Singh Rao
Anil Kumar Gupta

Active Tectonics of Kumaun and Garhwal Himalaya

 Springer

R. Jayangondaperumal
Structure and Tectonic Group
Wadia Institute of Himalayan Geology
Dehradun, Uttarakhand
India

V. C. Thakur
Structure and Tectonic Group
Wadia Institute of Himalayan Geology
Dehradun, Uttarakhand
India

V. Joevivek
Structure and Tectonic Group
Wadia Institute of Himalayan Geology
Dehradun, Uttarakhand
India

and

Akshaya College of Engineering and
    Technology
Kinathukadavu, Coimbatore, Tamil Nadu
India

Priyanka Singh Rao
Structure and Tectonic Group
Wadia Institute of Himalayan Geology
Dehradun, Uttarakhand
India

and

Geological Survey of India
Government of India
Hyderabad, Telangana
India

Anil Kumar Gupta
Structure and Tectonic Group
Wadia Institute of Himalayan Geology
Dehradun, Uttarakhand
India

and

Department of Geology and Geophysics
Indian Institute of Technology Kharagpur
Kharagpur, West Bengal
India

ISSN 2365-0656   ISSN 2365-0664   (electronic)
Springer Natural Hazards
ISBN 978-981-10-8242-9   ISBN 978-981-10-8243-6   (eBook)
https://doi.org/10.1007/978-981-10-8243-6

Library of Congress Control Number: 2018931503

Printed on acid-free paper

This Springer imprint is published by the registered company Springer Nature
Singapore Pte Ltd. part of Springer Nature
The registered company address is: 152 Beach Road, #21-01/04 Gateway East,
Singapore 189721, Singapore

*Dedicated to all those working on active fault
of the Himalaya*

Preface

In Himalaya, about one hundred thousand lives have been lost in large earthquakes during the past one hundred years. The loss of lives and property can be minimized through intervention of improved construction technology and incorporating building code. The Himalayan region is undergoing fast economic development on a large scale in the areas of hydroelectric power generation, building of highways including tunnels, tourism infrastructure, and other societal projects. In the frontal Sub-Himalaya and adjoining Ganga plain, there is ongoing development of industrial units and transport sector. It is in these types of infrastructure creation projects that the active fault studies (include mapping and characterization of a fault scarp using paleoseismological investigation) assume importance. A fault scarp is defined as a tectonic landform corresponding to, or roughly coincident, a fault plane that has displaced the ground surface. Genetic relationship between the active faults and earthquakes has now been established in the case of some of the destructive earthquakes, like 1994 Northridge, 1999 Chi-Chi, and 2005 Kashmir earthquakes. This has given rise to recognition of the hazard posed by active faults to the infrastructure and society. There are examples of near-total annihilation of a village, severe damage to educational infrastructure killing children and destruction of a large bridge over a river located close to the active faults in areas moderately damaged by earthquakes. In Himalaya, the major projects have been built taking into consideration the peak ground acceleration parameter. In the absence of an active fault in the project area, this may hold good and may be applicable for assessing vulnerability and safety of the infrastructure, whereas in the presence of an active fault in the vicinity of an infrastructure project, there will be an exponential increase in the risk factor. It is therefore imperative that the engineers also consider the occurrence or presence of an active fault and its characteristics in selecting the location and designing the infrastructure project. The active fault study has developed into an important area of research in earthquake hazard assessment, involving mapping, tectonic geomorphology, paleoseismology, and geochronological dating. The multidisciplinary approach is applied to determine the active fault properties that include the timing and magnitude of slip and vertical offset. There is now global recognition, and there are ongoing international programs on active faults. Kumaun and Garhwal are the two regions of Uttarakhand State. The state suffered loss of lives and property during the moderate 1991 Uttarkashi and 1999 Chamoli earthquakes. In the historical past, there was a major devastating earthquake of AD 1803 in Garhwal. There is also paleoseismic evidence that a great paleoearthquake in AD 1344 ruptured the Himalayan front and reactivated the Himalayan frontal fault. The large earthquakes originate in the Lesser Himalayan hinterland and the ruptures propagate to the forel and breaking the ground surface or remain blind at depths like the 2015 Gorkha earthquake in Nepal. The Sub-Himalaya zone between the Main Boundary Thrust and the Himalayan Frontal Thrust is characterized by active faults. In this book, we have focused our study on the active faults of this zone in Uttarakhand.

Dehradun, India   R. Jayangondaperumal
July 2017   V. C. Thakur
                                                                                                        V. Joevivek
                                                                                             Priyanka Singh Rao
                                                                                              Anil Kumar Gupta

Acknowledgements

Throughout the process of writing this book, many individuals from the Earth Science Community have taken time out to help us in this endeavor. We appreciate the community as a whole for actively participating in the feedback and contributions for this book. We express our gratitude to the Director, Wadia Institute of Himalayan Geology (WIHG), Dehradun, for partial financial support and providing the infrastructure and laboratory facilities through institutional flagship project to carry out this work. We would like to thank the Management of Akshaya College of Engineering and Technology, Coimbatore, for providing the infrastructure support to complete this task. We appreciate Mr. Rajeeb Lochan Mishra for helping us in the process of editing the document. The authors thank Prof. Robert S. Yeats and the anonymous reviewers for their useful comments in improving the book chapters. Thanks to publisher for accepting our proposal and publishing the book under Springer Natural Hazards book series. We would extend our gratitude to the editorial members for their assistance in editing, proofreading, and design of the book. Above all, we would like to thank our family members for their support.

About the Book

The book contributes to understanding the pattern of strain release and the level of seismic hazard imposed by large-great earthquakes in the frontal fold-thrust belts of Kumaun and Garhwal regions of Uttarakhand. The motivation for active fault studies and their characterization have been emphasized. The book presents the compilation of knowledge garnered in multidisciplinary or proxy studies involved in the understanding of seismic hazard in general and Kumaun–Garhwal Himalaya regions in particular with lucid new maps draped on modern Cartosat or SRTM DEM data. It also discusses satellite image calibration, active faults identifications, and map productions with flowchart. The book discusses window-wise active fault elements with attributes together with the tectonic geomorphic map. It also includes active fault scarp with topographic profile along with field photographs. Finally, it reviews all existing seismotectonic models of the Himalaya, its earthquake hazard, and its vulnerability, specifically for Kumaun and Garhwal regions.

Contents

About the Authors

Dr. R. Jayangondaperumal is a Senior Scientist in the Wadia Institute of Himalayan Geology, Dehradun, India. He has been working in research projects related to structural geology, earthquake geology, and tectonics for over one and a half decade. He and his team have excavated about 25 trenches for paleoseismological study along the Himalayan frontal zone. He has significantly contributed in understanding the relationship between strain accumulation and release in the context of mechanics of earthquakes and mountain building processes so as to quantify the seismic hazard associated with the Himalayan continental collision. He has also carried out seismic hazard assessment for the Subansiri hydropower project in the NE Himalaya. He has published over 46 papers and contributed 2 chapters and 1 Indian patent. Currently, he along with his large team of students is carrying out national-level research projects on earthquake geology in NE and NW Himalaya.

Dr. V. C. Thakur is M.Sc. (Panjab University, India), and Ph.D. (DIC, London University). He joined the Wadia Institute of Himalayan Geology in 1972. He has worked extensively in unraveling the regional framework and tectonics of the remote regions of Ladakh, Zanskar, Chamba, Garhwal, Kumaun, and Arunachal. On his retirement as Director of the Institute, he was awarded Emeritus Scientist of CSIR; since then he has been working in several research projects on active tectonics of the Himalayan frontal zone. He has published more than 130 papers and authored a book 'Geology of Western Himalaya,' Pergamon Press, Oxford. He is a Fellow of the Indian Academy of Sciences (1991), recipients of the National Mineral Award (1984) and Padma Shri Award (2018).

Dr. V. Joevivek is a Professor and Dean (Research) at Akshaya College of Engineering and Technology, Kinathukadavu, Coimbatore, India. His field of specialization is remote sensing and physical geography. He has published over 18 papers and 2 chapters and contributed 2 Indian patents. He is a Fellow of the Society of Earth Scientists and member of various scientific societies. This book is part of his research work carried out at the WIHG under institutional flagship project.

Ms. Priyanka Singh Rao is a Geologist working in Geological Survey of India and also pursuing her Ph.D. degree in 'Surface Rupture Investigations of the 1950-Meisoseismal Zone of Assam Earthquake along Himalayan Foothill Zone, Arunachal Himalaya.' She discovered surface faulting of 1950 Assam earthquake along the eastern Himalayan Frontal Thrust.

Prof. Anil Kumar Gupta is a Faculty at the Indian Institute of Technology Kharagpur, Kharagpur, India. Previously, he was the Director of the Wadia Institute of Himalayan Geology, Dehradun (2011–2017). He has extensively worked to study paleoceanographic and paleoclimatic evolution of the Indian Ocean with special reference to the understanding of short- and long-term changes in the Asian/Indian monsoons and their connections with the Himalayan uplift. His research has been widely cited and recognized through several prestigious awards including the Indian Science Congress Association Young Scientist Award 1990, the National Mineral Award (2000), and the TWAS Prize in Earth Sciences (2010). He is the Fellow of several science academies (INSA, NASI, IASC, TWAS) and is the recipient of J.C. Bose Fellowship.

Chapter 1
Introduction

The Himalaya constitutes one of the highest mountains of the world with 37 mountain peaks rising ~ 7000 m above the mean sea level (MSL). It is an orogenic belt of the Cenozoic Era that extends laterally from west to east for ~ 2500 km in a curvilinear framework with northwest and northeast syntaxial bends on its western and eastern extremities, respectively. The Himalaya is divided from south to north into four parallel longitudinal mountain belts of varying width, each having its own physiographic features and geological history. They are the outer or Sub-Himalaya, the Lesser or Lower Himalaya, the Higher or Greater Himalaya, and the Tethys or Tibetan Himalaya. Farther north lies the Trans Himalaya which includes the Ladakh, Karakoram, and Mansarovar ranges. The Himalaya lies between the Indian subcontinent to the south and the high plateau of Tibet with ~ 4 km average altitude to the north. The Tethys Himalaya and largely the Greater Himalaya constitute the southern margin of the Tibetan Plateau in physiographic framework. There is a distinct physiographic transition from the Great Himalaya to the Lesser Himalaya.

   The Himalaya represents a convergent plate boundary between the Indian and the Asian plates. Seismicity in the Himalaya is a consequence of the ongoing northward convergence of India and the crustal deformation. A fault scarp is defined as tectonic landforms corresponding, or roughly coincident, with a fault plane that has displaced the ground surface (Nakata and Kumahara 2002; Kaneda et al. 2008) due to earthquake. There is an inherent relationship between active faults and earthquakes. The role of active faults in the occurrence of highly destructive earthquakes has now been established, for example, 1994 Mw ~ 6.7 Northridge (Stein et al. 1994), 1999 Mw ~ 7.6 Chi-Chi (Yu et al. 2004), 2005 Mw ~ 7.6 Kashmir (Avouac et al. 2006; Jayangondaperumal and Thakur 2008; Kaneda et al. 2008; Thakur et al. 2010), and 2015 Mw ~ 7.8 Gorkha (Avouac et al. 2015; Parameswaran et al. 2015; Gualandi et al. 2016). This has led to the growing recognition of the hazard posed by active faults (i.e., fault scarp) to the infrastructure and society, highlighting the need to understand the association between active fault parameters and the earthquakes.

© Springer Nature Singapore Pte Ltd. 2018
R. Jayangondaperumal et al., *Active Tectonics of Kumaun and Garhwal Himalaya*,
Springer Natural Hazards, https://doi.org/10.1007/978-981-10-8243-6_1

Five major earthquakes have occurred in the Himalayan region during the last hundred years (Fig. 1.1): 1905 Kangra Mw ∼7.8 (Middlemiss 1910; Ambraseys and Bilham 2000), 1934 Bihar–Nepal Mw ∼8.2 (Dunn et al. 1939; Bilham 1995), 1950 Assam (now Arunachal) Mw ∼8.6 (Chen and Molnar 1990; Priyanka et al. 2017), Kashmir 2005 Mw ∼7.6 (Hussain et al. 2009), and 2015 Gorkha magnitude Mw ∼7.8 (Avouac et al. 2015). These earthquakes occurred over the detachment and their ruptures were distributed between the Himalayan Frontal Thrust (HFT) and the Main Central Thrust (MCT) (Seeber and Armbruster 1981; Mugnier et al. 2013). The detachment represents the basal decollement which is now referred as the Main Himalayan Thrust (MHT). The major earthquakes have brought great devastation in terms of loss of lives and property. The 1905 Kangra event killed 14,000 people, and about 80,000 persons perished in the 2005 Kashmir earthquake. Moderate earthquakes like the 1991 Uttarkashi and 1995 Chamoli of moment magnitudes Mw 6–6.5 also inflicted considerable damage to life and property in Garhwal.

Pioneering studies of active tectonics have been carried out in the Himalaya by Nakata (1972, 1975) in the foothills of Darjeeling Himalaya, Pinjore, and Dehradun areas and later by Valdiya (1992a, b, and references therein) in the Kumaun Himalaya. The Geological Survey of India (GSI) carried out neotectonic studies in the Riasi, Katra, and the adjoining areas, as a part of their geotechnical study for construction of the Salal hydroelectric project across the Chenab River (Krishnaswamy et al. 1970; Nawani et al. 1982). These authors described reactivation of the Riasi Thrust as late as Pleistocene and Holocene, based on their field observations and mapping. In the recent years, reactivation of the Riasi Thrust has been investigated more extensively by several workers (Thakur et al. 2010; Vassallo et al. 2015; Gavillot et al. 2016). Studies of active tectonics have been focused mainly on the Himalayan frontal zone. Tectonic geomorphological investigations including active fault mapping have been conducted in the frontal zone between the MBT and the HFT (Philip and Virdi 2006; Thakur et al. 2007; Philip et al. 2011, 2012; Jayangondaperumal et al. 2010a, b, 2017a, b; Malik et al. 2010a, b, 2016, and others). Paleoseismological studies have been undertaken along the HFT in the northwest Himalaya indicating surface ruptures of historical earthquakes (Kumar et al. 2001, 2006, 2010; Malik and Mathew 2005, Malik et al. 2010a, b; Philip et al. 2011; Jayangondaperumal et al. 2011, 2013, 2017a; Mishra et al. 2016; Priyanaka et al. 2017). Shortening and slip rates have been estimated on the HFT using geomorphic surfaces (Wesnousky et al. 1999; Kumar et al. 2006; Malik et al. 2010b; Jayangondaperumal et al. 2013; Thakur et al. 2014). The ongoing horizontal velocity rates have been estimated using the global positioning system (GPS) across the northwest Himalaya (Banerjee and Burgmann 2002; Mullick et al. 2009; Jade et al. 2007, 2011; Ponraj et al. 2011).

During the past decade, a wealth of new data—geologic, paleoseismic, and geodetic—unequivocally reveals that the Himalayan basal decollement (i.e., the Main Himalayan Thrust, MHT) (Zhao et al. 1993; Nelson et al. 1996; Nabelek and HI-ClIMB Team 2009) sporadically ruptures in very large earthquakes, some of which break the surface of mountain front (Lave et al. 2005; Kumar et al. 2006,

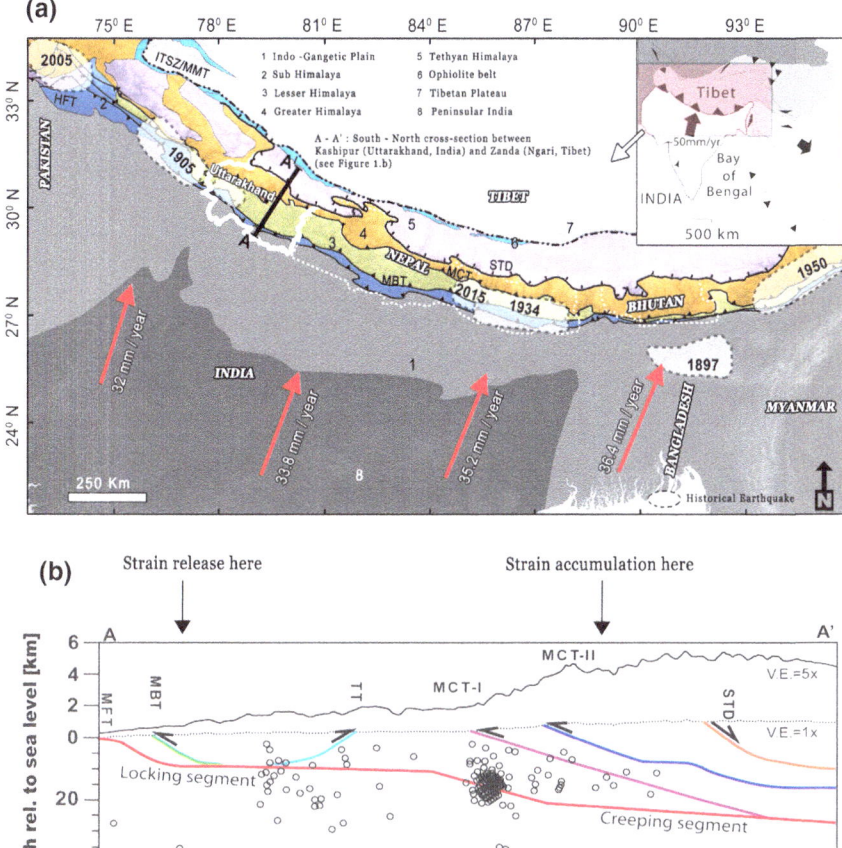

Fig. 1.1 **a** Outline tectonic map of Himalaya showing regional framework, principal tectonic zones, and rupture areas of major earthquakes. Red arrows point to convergence direction and estimated rate of convergence derived through GPS measurements. Inset white rectangle with irregular boundary is the location of the study area—Uttarakhand State. **b** S–N cross section across Himalaya demonstrating principal tectonic zones and faults. This seismotectonic model depicts the locked and ductile creeping segments of the MHT and zones of strain accumulation and release (compiled and modified from Caldwell et al. 2013 and Rawat et al. 2014). Location of the cross section is shown as S–N line in the map above

2010; Malik et al. 2010a, b; Jayangondaperumal et al. 2011, 2017a; Kumahara and Jayangondaperumal 2013; Sapkota et al. 2013; Bollinger et al. 2014; Rajendran et al. 2015; Mishra et al. 2016; Priyanka et al. 2017). These observations suggest that the active shortening across the Himalayan wedge is mostly accommodated by slip along the narrow Sub-Himalayan zone (i.e., between MBT and HFT) and its linked decollement beneath the Himalaya. Between earthquakes (i.e., in the interseismic period), the MHT is locked, and strain builds across the transition between the locked portion and the down-dip ductile regime with aseismic slip (Thatcher 1983). The accumulated interseismic strain relieves through earthquakes and results in slip across the up-dip fault reach (Stein et al. 1988).

Earthquakes in the fold-and-thrust belt of the Himalaya occur both on the basal decollement (Seeber and Armbruster 1981; Mugnier et al. 2013) and on faults in the upper plate that break the ground surface as surface rupture in the case of 2005 Kashmir earthquake (Kaneda et al. 2008) or remain blind within the subsurface depth, e.g., Gorkha 2015 earthquake. Earthquake geology thus provides unique insight into the processes that allow continents to collide and the associated seismic hazards (Yeats et al. 1992; Bilham et al. 2001). Active faults are associated with the earthquakes. Earthquake surveys in many parts of the world have demonstrated that the maximum damage to infrastructure and property occurred on or close to the surface rupture of the earthquakes.

Faults are commonly considered to be active if they have moved one or more times in the last 10,000 years (USGS Glossary). An active fault is likely to become the source of another earthquake sometime in the future. A fault is considered active if there has been movement observed or evidence of seismic activity during the last 10,000 years. Faults recognized at the surface having evidence of displacement in the Quaternary period with a potential of generating earthquakes are also categorized under active faults. Faults that have not moved during the Quaternary period (2.5 Ma) are generally classified as inactive (Keller and Pinter 1999). Mapping of active faults and their characterization will provide the first direct constraints in a region known to have earthquakes. Short- and long-term deformation rates of these active faults are essential because surface rupture only accompanies earthquakes larger than a given size (Wells and Coppersmith 1994). By characterizing slip per event for active faults at the Himalayan front and within the wedge, we can explore moment release across the Himalaya. These data allow the addressing of key questions such as what is the recurrence of mega-thrust, subduction like events on the MHT and how and at what rate is strain release distributed across the Himalaya? The value of these data will provide accurate input for seismic hazard as the region has one of the highest slip deficits along the Himalaya (Bilham et al. 2001).

Uttarakhand State lies between Nepal to the east and Himachal Pradesh to the west and encompasses Garhwal and Kumaun Himalaya. Yamuna, Ganga, Ramganga, and Goriganga in Uttarakhand and Kali between Nepal and Uttarakhand are the major Himalayan Rivers having great potential for hydroelectric power generation. Uttarakhand is undergoing economic transformation through infrastructure and industrial development. The understanding of active faults and earthquakes in the region is relevant and important for constructing the safer earthquake-resistant infrastructure projects, housing and urban development projects. Landslides are another natural hazard in the mountains. Their frequent occurrence causes damage to property, disruption to road traffic, and loss of lives. In some cases, there is a linkage between the landslides and active faults. The population density in the Himalaya including Uttarakhand increases from the High,

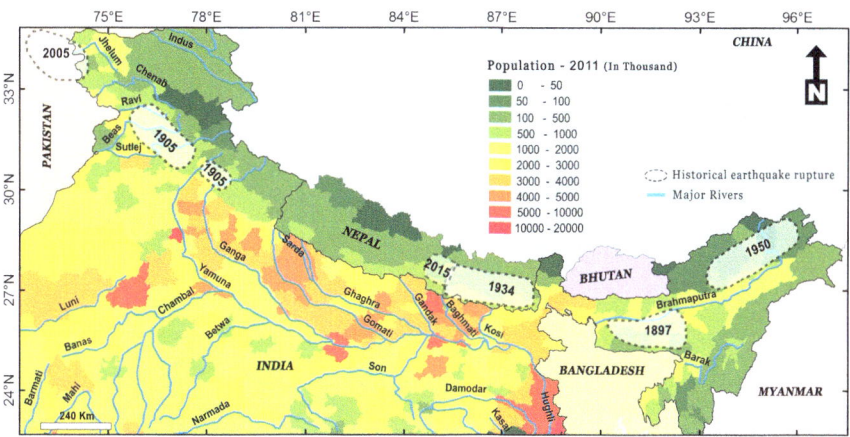

Fig. 1.2 Map depicting population density in the Himalaya and the northern part of Indian subcontinent, together with rupture zones of major earthquakes. In the Himalaya and Indo-Gangetic Plains, the population density increases from the Greater to the Sub-Himalaya and the Indo-Gangetic Plains. Strain released through earthquakes affects largely the lesser and the Sub-Himalaya. Earthquake surface ruptures reported from the Himalayan front (HFT) also pose earthquake hazard to the Indo-Gangetic Plains. District-wise population data extracted from http://censusindia.gov.in/ (for India) and population atlas of Nepal, 2014, 194p (for Nepal)

Lesser, Sub-Himalaya to Gangetic Plains (Fig. 1.2). The intensity records of past earthquakes reveal that the major damage was confined to the Lesser and Sub-Himalayan zones. However, paleoseismological investigations show that the ruptures of major earthquake propagated to the foreland and broke the ground surfaces on the Sub-Himalayan front adjoining the alluvial plains. This poses serious earthquake hazard and vulnerability to the population and infrastructures in the alluvial plain areas. Uttarakhand State has a recorded history of 1803 Garhwal earthquake of magnitude Mw 7.6–8 and moderate magnitude, Mw ∼6, Uttarkashi and Chamoli earthquakes. These earthquakes are supposed to have originated over the basal decollement called the MHT or related to one of the faults emerging from the Himalayan wedge.

1.1 Himalayan Tectonics

In tectonic framework, the Himalaya is divided into five tectonic zones from south to north extending longitudinally along its entire length (Fig. 1.1). The southernmost Sub-Himalaya zone abuts against the Indo-Gangetic Alluvial Plain (IGAP) along a tectonic boundary called the Main Frontal Thrust (HFT) or the Himalaya Frontal Thrust (HFT).

The Sub-Himalaya represents the foreland basin constituting marine Paleocene–middle Eocene and fluvial Miocene–Pleistocene sediments. The Lesser Himalaya

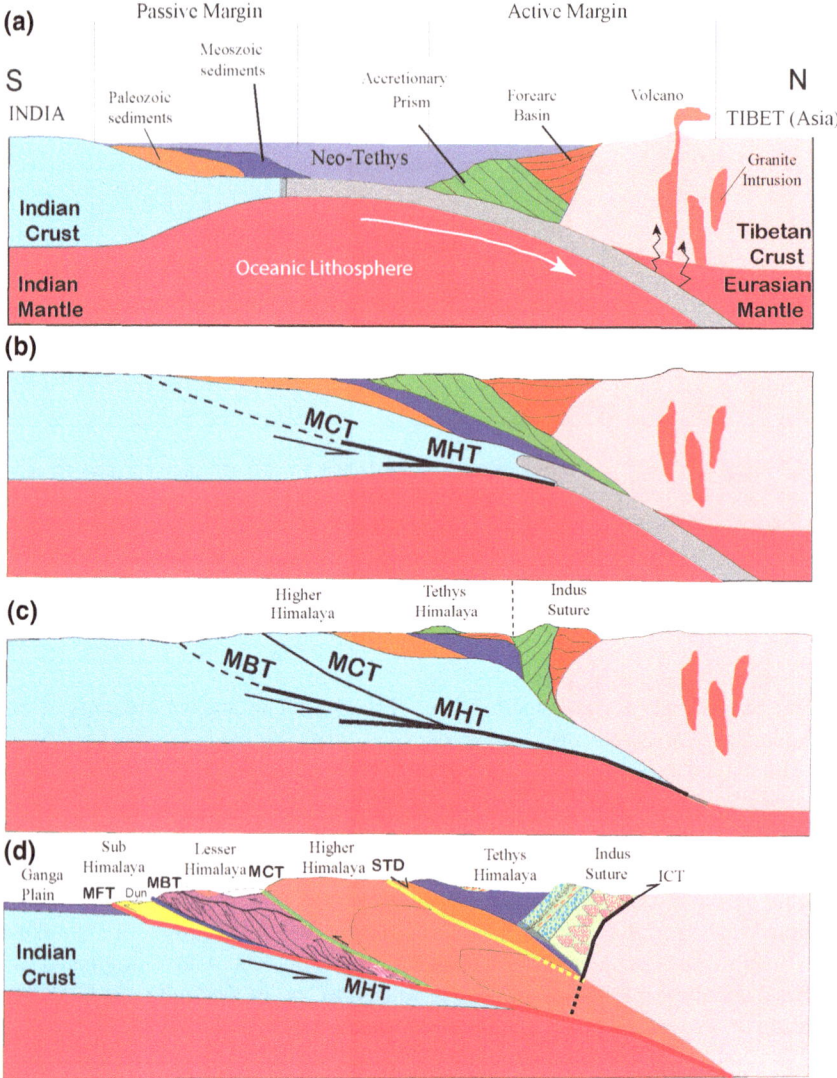

zone is separated to the south from the Sub-Himalaya along the Main Boundary Thrust (MBT). The Lesser Himalaya is made of late Proterozoic–lower Cambrian sediments, originally belonging to the northern part of Indian continent, overlain by the crystalline thrust sheets. The High Himalaya zone overrides the Lesser Himalaya along the Main Central Thrust (MCT) characterized by a shear zone. The High Himalaya is the main metamorphic belt comprising of green schist to amphibolites rocks with Miocene granites. It is overlain by the Tethys Himalaya zone along the normal fault system of the South Tibet Detachment (STD). The Tethys Himalaya is composed of late Precambrian to Cretaceous–Lower Eocene

◄**Fig. 1.3** Schematic diagram from the top to bottom shows tectonic evaluation of the Himalaya.
a Northward movement of India with respect to Tibet leads to subduction of the oceanic
lithosphere of the Indian plate under Tibet (Asia). The subducting Indian oceanic lithosphere at
greater depth produced magma due to partial melting, and the uprising magma generated plutonic–
volcanic arc. At the junction between the oceanic plate and Tibet continent, accretionary wedge
was developed, and the eroded flux was deposited in the fore-arc basin. **b** Collision occurred
around 50 Ma between the continents of India and Tibet. The shortening induced through collision
in the Indian crust gave rise to development of the Main Central Thrust (MCT) during early
Miocene, as emerging from the Main Himalayan Thrust (MHT). **c** The ongoing convergence of
India initiated formation of the Main Boundary Thrust (MBT) south of the MCT and emerging
from the MHT during middle Miocene. **d** The major fold-thrust system was developed between the
MCT and MBT by Pliocene. Foreland basin was developed and the ongoing convergence
produced the foreland propagating fold-thrust system with the emergence of the MHT as the HFT
(MFT) on the Himalayan front (compiled and modified from Molnar 1986)

marine sequence deposited in the Tethys Sea. The Tethys Himalaya zone represents
the northern boundary of the Indian Plate and separates the Trans Himalaya of the
Asian Plate along a tectonic boundary, referred as the Indus/Tsangpo/Main Mantle
Thrust. There was a vast Neo-Tethys Ocean between the Indian and Asian conti-
nents in Mesozoic times. The opening of central Indian Ocean ridge resulted in N–
NE movement of the Indian plate and that leads to closing of the Neo-Tethys Ocean
(Fig. 1.3a).

The Mesozoic sequence of the Tethys Himalaya constituted the north-facing
passive margin of the Indian plate. Toward north of the passive margin, the oceanic
Indian plate subducted beneath the Asian plate, leads to the closing of the
Neo-Tethys Ocean (Fig. 1.3b). The Trans Himalaya zone, an active margin of the
Asian plate, is characterized by the petrotectonic assemblages that reveal the
paleotectonic regimes during the closing of the NeoTehys (Searle et al. 1987;
Hodges 2000; Yin 2006). Along the Indus–Tsangpo Suture zone, the ophiolitic
melanges, dismembered ophiolites, and deep oceanic deposits of Jurassic age are
the oceanic crust remnants of the closed Neo-Tethys. The occurrence of Cretaceous
island—arc volcanics of Dras and Kohistan and Andean—type granite batholiths of
Ladakh, Deosai and Gangdese were generated over the Asian plate as a result of
subduction of the oceanic Indian plate beneath the Asian plate (Thakur 1992;
Hodges 2000; Yin 2006). There was no deformation of the Indian plate during the
subduction, as the plate was moving passively beneath the Asian plate. The dis-
appearance of Tethys Sea is marked by transgression–regression of the last marine
Paleocene–Eocene sediments. It is coincident with the timing of collision around
50 Ma of the Indian plate with the Asian plate. Post-syn-collision, northward
convergence of India produced crustal shortening of the Indian plate (Fig. 1.3). The
thickening of the crust due to crustal scale imbrication through accretion was
responsible for the formation of High Himalayan metamorphics and the MCT
(Fig. 1.3c). The MCT was developed in middle Miocene and responsible for
exhumation of the High Himalaya.

However in channel flow model, the middle crust of south Tibet was extruded
into the High Himalaya facilitated by the movements over the MCT and the STD
(Beaumont et al. 2001; Grujic et al. 2006). The Lesser Himalaya represented the

(a)

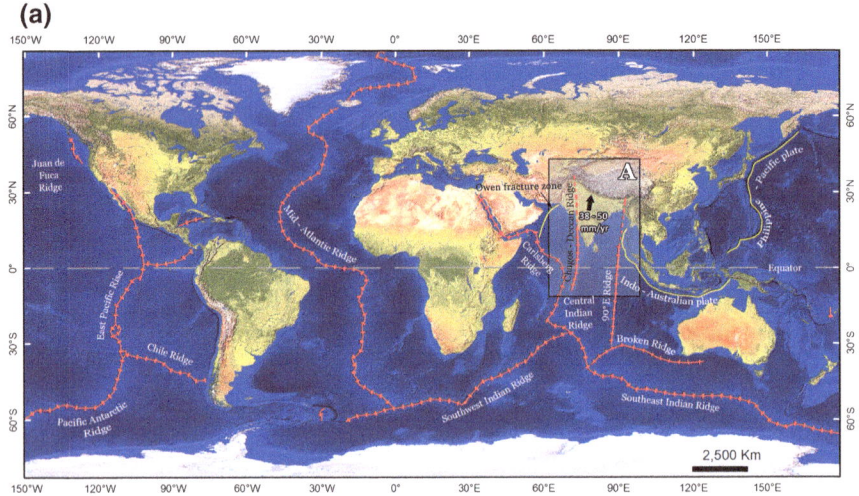

Fig. 1.4 a Global map showing oceanic ridges (red line with cut) which represent the divergent plate boundaries and opening of the ocean. Indian Ocean is opening along the NNW trending Central Indian Ridge as a divergent plate boundary between the Indian and African plates at a rate of 38–50 mm/year. This results in N–NE convergence of the Indian plate. Inset 'A' explained in detail in Fig. 1.4b. **b** Detailed schematic diagram of inset A shown in Fig. 1.4a demarcates tectonic and structural trends of the Indian subcontinent

northern part of the Indian continent which was deformed in late Miocene–Pliocene with the formation of a thrust duplex system between the MCT and the MBT (Srivastava and Mitra 1994; Celerier et al. 2009). South of the MBT, Sub-Himalayan foreland basin was developed with deposition of marine Paleocene–middle Eocene and the fluvial sediments of Miocene–Pleistocene age (Fig. 1.3d). The subsidence of the foreland basin was brought about by bending of the Indian plate due to the load of southward advancing thrust sheets (Lyon-Caen and Molnar 1985).

1.2 The Basal Decollement (MHT) is an Active Fault

The collision between the Indian and Asian plates and north-northeast convergence of the Indian plate with respect to the Asian plate resulted due to opening of the Indian Ocean along the Central Indian Ridge at a rate 38–50 mm/year (Fig. 1.4). The seismicity in the Himalaya is consequence of this active convergence tectonics. The outstanding feature of Himalayan seismicity is that most of the epicenters lie along a relatively narrow belt between the MBT and the MCT (Fig. 1.5). In the central Himalaya including Nepal, Kumaun, and Garhwal, the seismicity is concentrated in a ∼30-km-wide belt at a depth of 10–20 km beneath the topographic front of the High Himalaya encompassing the MCT zone (Ni and Barazangi 1984;

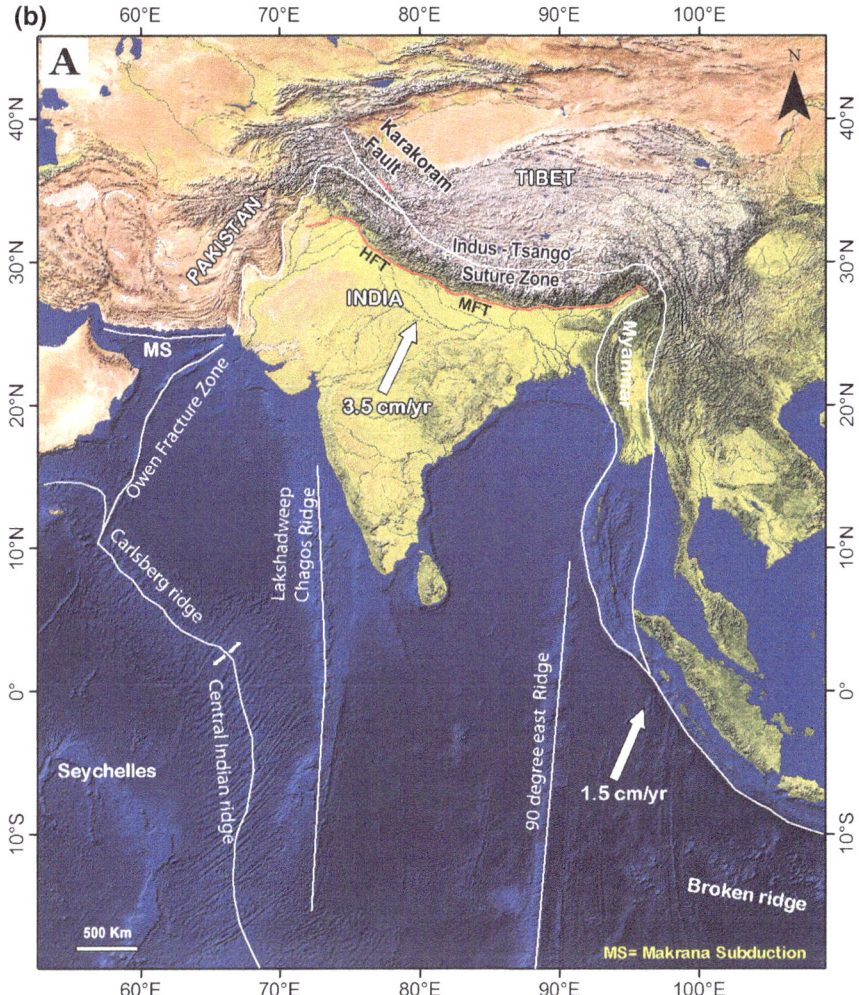

Fig. 1.4 (continued)

Pandey et al. 1995; Arora et al. 2012), whereas in the northwest Himalaya, seismicity zone lies much south of the High Himalaya (Armbruster et al. 1978; Naresh et al. 2009) (Fig. 1.5).

Northward converging India against Asia produced crustal shortening of the northern margin of the Indian continent through imbrications of the Indian crust along the major intra-crustal thrusts. Stacking of thrust slabs one upon another through accretion gave the overall architecture of the Himalaya (LeFort 1975; Molnar 1984; Hodges 2000; Yin 2006). The major intra-crustal thrust fault systems in the central Himalaya are the Main Central Thrust (MCT), Munsiari–Ramgarh Thrust, the Main Boundary Thrust (MBT), the Lesser Himalaya duplex, and the

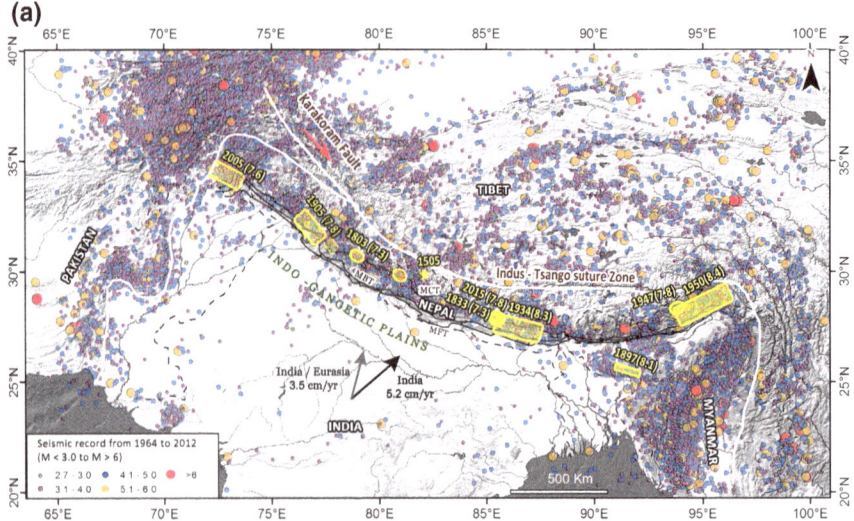

Fig. 1.5 **a** Outline map of Himalaya showing instrumentally recorded microseismicity along with major earthquakes (red color). Also showing extensional graben structures in Southern Tibet shows east–west extension, whereas convergence is dominated by thrust fault mechanism. **b** Outline map of Himalaya showing Himalayan Frontal Thrust (HFT/HFF), Main Boundary Thrust (MBT), Main Central Thrust (MCT), and Indus–Tsangpo Suture (ITS) together with trenched fault zone location (Square box) along Himalayan Arc. The earthquake epicenters are largely distributed between the MCT and the MBT. Square box with numeral (1–30) shows location of trench excavated across the fault scarp for paleoseismological investigation by various researchers. Trench location and the corresponding literature are listed as follows: 1: Muzaffarabad-Kondo et al. (2008), 2: Riasi-Vassallo et al. (2012), 3: Hajipur-Malik et al. (2010a, b), 4: Bhatpur-Kumahara and Jayangondaperumal (2013), 5: Mehandpur-Jayangondaperumal et al. (2017a), 6: Chandigarh-Malik et al. (2008); Kumar et al. (2010), 7: Kala Amb-Kumar et al. (2001), 8: Rampur Ganda-Kumar et al. (2006), 9: Dehradun-Kumar et al. (2006), 10: Laldhang-Kumar et al. (2006), 11: Ramnagar-Kumar et al. (2006), 12: Belparao-Rajendran et al. (2015), 13: Ramnagar-Malik et al. (2016), 14: Mohana Khola-Yule et al. (2006), 15: Bandelpokhari-Upreti et al. (2008), 16: Butwal-Sapkota and Rimal (1997), 17: Aurahi-Upreti et al. (2000), 18: Mahra Khola-Lave et al. (2005), 19: Sir Khola-Sapkota et al. (2013), 20: Thapatol-Bollinger et al. (2014), 21: Tokla-Nakata et al. (1998), 22: Hokse-Nakata et al. (1998), 23: Chalsa-Kumar et al. (2010), Panijhora-Mishra et al. (2016), 24: Nameri-Kumar et al. (2010), 25: Harmutty-Kumar et al. (2010), 26: Himebasti-Jayangondaperumal et al. (NHPC Report, 2011), 27: Nigluk T2 (Paper in preparation), 28: Nigluk T1 (Paper in preparation), 29: Marbang-Jayangondaperumal et al. (2011), and 30: Pasighat-Priyanka et al. (2017)

Himalayan Frontal Thrust (HFT) (Robinson et al. 2006; Robinson and Pearson 2013). This structural–tectonic framework extends to the west in Kumaun and Garhwal Himalaya.

The major thrust faults continue longitudinally along the entire length of the Himalaya. The principal thrusts were developed progressively from north to south, the MCT in early Miocene (Hubbard and Harrison 1989), the MBT in late Miocene (Meigs et al. 1995; Jayangondaperumal and Dubey 2001), and the HFT in Quaternary (Mugnier et al. 2005; Thakur et al. 2007; Jayangondaperumal et al.

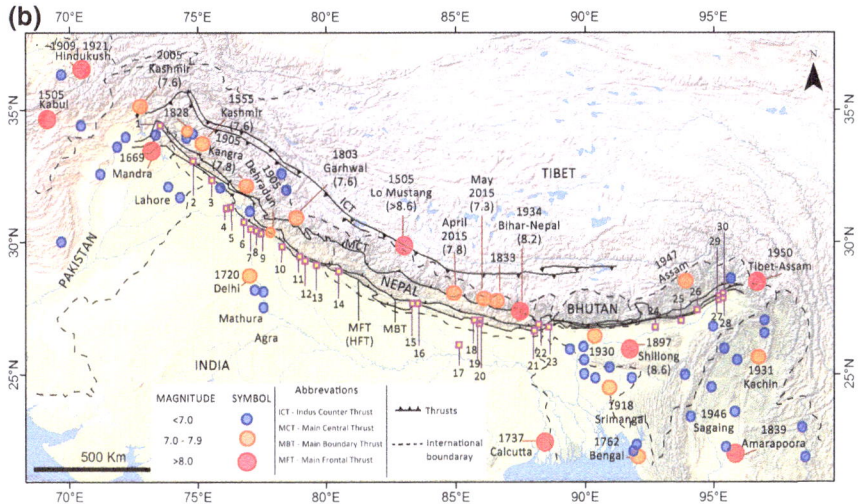

Fig. 1.5 (continued)

2010a, b). The principal thrusts sole at variable depth with the basal decollement designated as the Main Himalayan Thrust (MHT) (Fig. 1.1b). The existence of the MHT is established through geophysical experiments INDEPTH and HIMNET with combination of geophysical data (Brown et al. 1996; Hauck et al. 1998; Nabelek and HI-ClIMB Team 2009) and from balanced cross sections prepared along mapped corridors across the Himalaya (Schelling and Arita 1991; Schelling 1992; Srivastava and Mitra 1994; DeCelles et al. 1998, 2001).

The MHT extends at shallow dip (2°–6°) from depths of 4–5 km beneath the foreland to depths in excess of 40 km beneath southern Tibet (Schelling 1992; Power et al. 1998; Hauck et al. 1998; Mugnier et al. 1999; Nabelek and HI-ClIMB Team 2009). The MHT represents the plate boundary between the Indian and Asian plates. The southward growing Himalaya progressively shifted toward the south along the MHT.

The Himalayan Frontal Thrust (HFT), also referred as the Main Frontal Thrust (MFT), represents the surface expression of the outer ramp part of the MHT. The MFT is marked at the surface by a series of isolated fault-related folds (Yeats et al. 1992; Lave and Avouac 2000; Champel et al. 2002; Valdiya 2003; Mugnier et al. 2004; Delcaillau et al. 2006; Yeats and Thakur 2008). Each fault–fold pair represents an individual imbricate fault splay from the MHT (Schelling 1992; Power et al. 1998; Mugnier et al. 2004). A south-vergent thrust fault cuts the southern fold limb of nearly every fold in the central Himalaya from Nepal to India (Wesnousky et al. 1999; Valdiya 2003; Mugnier et al. 2004; Thakur and Pandey 2004; Lave et al. 2005; Kumar et al. 2010; Malik et al. 2010a, b; Jayangondaperumal et al. 2017a, b). In the Kangra reentrant, the frontal thrust demonstrably emerges at the surface cuts the forelimb of the Janauri anticline to the southeast of the Beas River (Malik et al. 2010a, b; Kumahara and

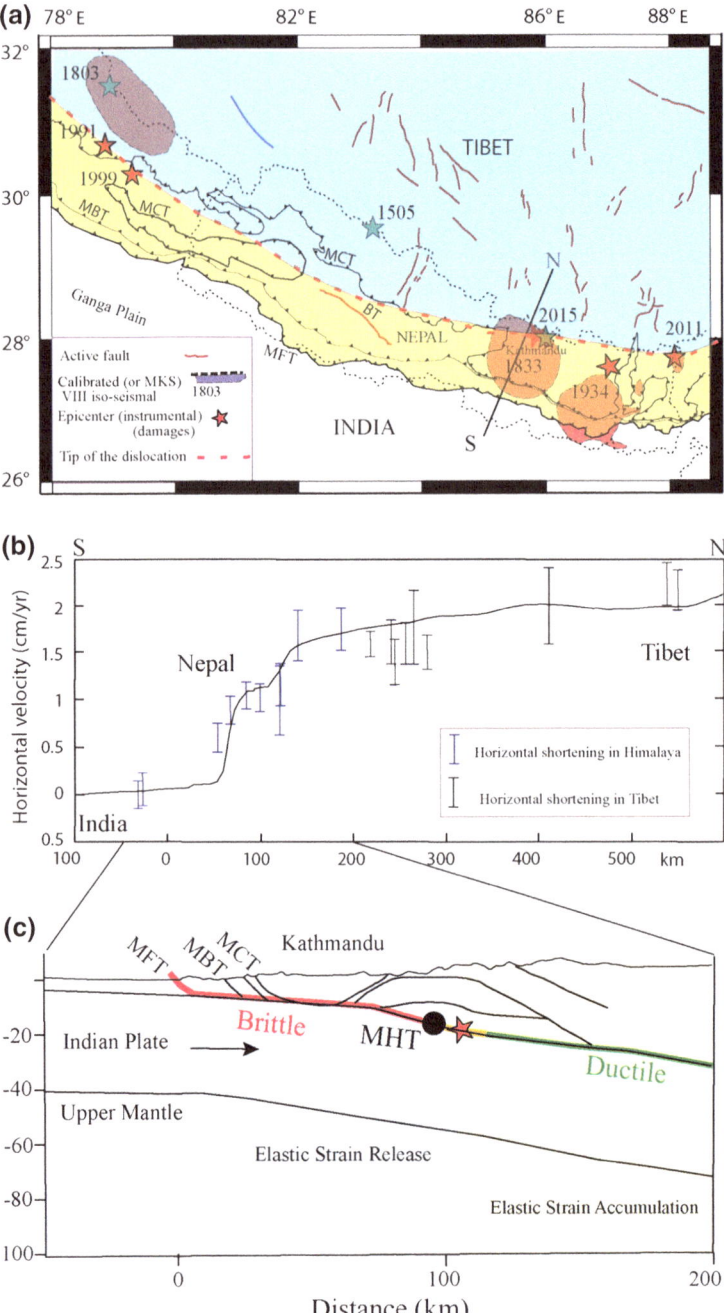

◄**Fig. 1.6   a** A simplified map showing locking line (at ∼ base of the High Himalaya) separating a region of elastic strain accumulation to the north (blue) from the zone of elastic strain release to the south (yellow) along with major tectonic elements. **b** A cross section across central Himalaya in Nepal showing major regional structures. The Main Himalayan Thrust (MHT), with ramp and flats, depicts brittle and ductile regimes (Modified from Mugnier et al. 2013). **c** The brittle and ductile portions of the MHT are characterized by strain release and strain accumulation, respectively, modeled based on GPS velocity

Jayangondaperumal 2013; Jayangondaperumal et al. 2017a, b). Along strike to the NW in Kashmir Himalaya, the Surin-Mastgarh anticline (SMA) marks the deformation front, but the forelimb of SMA is not cut by a fault at the surface according to existing mapping (Raiverman et al. 1994).

The tectonic–structural framework of the Himalaya indicates principally compressional tectonics, and the earthquakes largely invoke reverse (thrust) fault mechanism with strike-slip component increasing toward northwest and eastern Himalaya. The geodetic (GPS) measurements imply N–NE convergence and building of elastic strain during interseismic interval and release of strain during earthquake events (Fig. 1.6). The rate at which strain accumulates and is released along major fault systems within the Himalaya, and how this behavior varies along strike of the Himalayan arc, remain enigmatic. The Himalaya grows during an earthquake (Fig. 1.7). Earthquakes incrementally translate the upper plate toward the foreland, accrete material to the toe and base of the wedge, and cause internal deformation (Bombolakis 1986, 1994; De Bremaecker 1987; Price 1981). Fault geometry and loading rate dictate deformation on earthquake timescales (Ader et al. 2012).

The geologically estimated convergence rate recorded on the Himalayan front decreases from the central Himalaya to the northwest Himalaya and increases from central to eastern Himalaya. It is 18–20 mm/year in Nepal (Lave and Avouac 2000), 13–15 mm/year in Garhwal and Kashmir (Wesnousky et al. 1999; Vassallo et al. 2015; Gavillot et al. 2016), and 23 mm/year in eastern Himalaya (Burges et al. 2012). In central Himalaya, Nepal, the shortening recorded on the MFT (or HFT) consumes nearly the entire slip on the Main Himalayan Thrust (MHT) (Lave and Avouac 2000), whereas in the northwest Himalaya, Kangra reentrant and Jammu area (Thakur et al. 2014; Vassallo et al. 2015; Gavillot et al. 2016), and Kameng in eastern Himalaya (Burges et al. 2012), the shortening is distributed across several active structures in the Sub-Himalaya.

The geodetic (GPS) measurements indicate that the Indian plate is underthrusting Tibet at a convergence rate of 45–51 mm/year (Gahalaut and Chander 1999; Bettinelli et al. 2006; Gahalaut and Kundu 2011). Of the total convergence, 18–20 mm/year is consumed in the Himalaya, and the remaining amount is accommodated in Tibet and farther north in Asia (Armijo et al. 1986; Avouac and Tapponnier 1993; Peltzer and Saucier 1996). The GPS-derived convergence rate varies from the central to the northwest Himalaya. The convergence rate is 18–20 mm/year in central Himalaya, Nepal (Bilham et al. 1997; Bettinelli et al. 2006; Ader et al. 2012), 13–15 mm/year in Garhwal and Himachal (Banerjee and Burgmann 2002; Ponraj et al. 2011), and 11–14 in Kashmir in NW Himalaya (Schiffman et al. 2013). Models of geodetic data imply that ∼110 km segment from the Himalayan front to the north is locked invoking very little convergence with plate boundary at shallow depth in brittle regime (Fig. 1.6).

The main convergence is largely consumed aseismically through ductile creep wherein the plate descends to greater depth of 18–20 km in ductile regime. The small circle arc describes the locking line on the MHT in space (Bilham et al. 2001; Banerjee and Burgmann 2002; Bettinelli et al. 2006; Feldl and Bilham 2006; Ader et al. 2012). The position of the locking line corresponds closely with the Himalayan topographic arc defined by the step between south Tibet and the Lesser Himalaya (Fielding 2000; Seeber and Gornitz 1983). This geometry indicates that a 'locking line' separates the up-dip locked and down-dip continuously slipping portions of the MHT (Figs. 1.7 and 1.8) (Bilham et al. 2001; Ader et al. 2012).

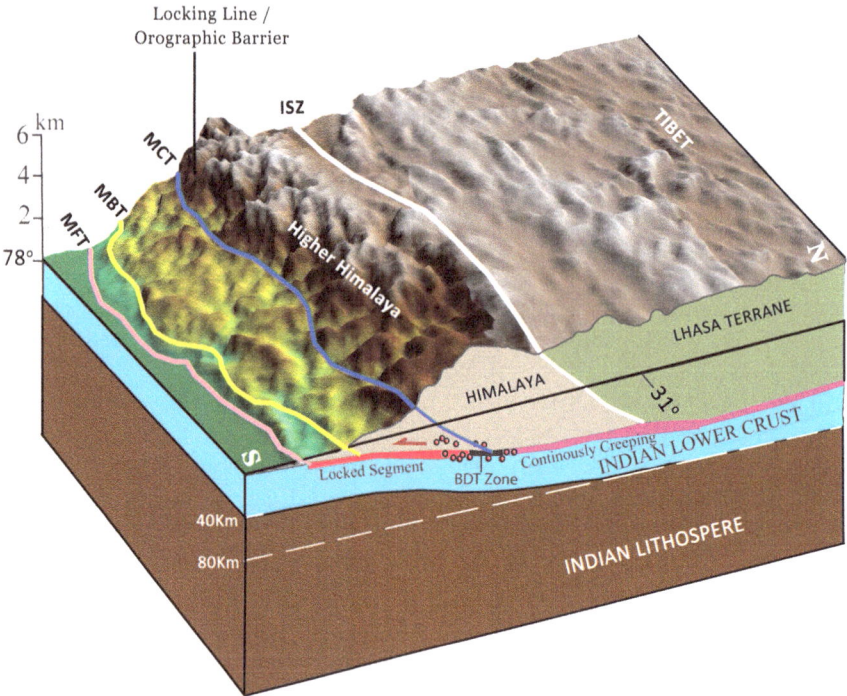

Fig. 1.7 Three-dimensional perspective view of seismotectonic model of Himalaya demonstrates underthrusting of Indian plate beneath the Lhasa terrain of Asian plate and ongoing northward convergence of India against Asia. GPS derived locked portion (red) and ductile creeping segment of the MHT (green). Solid red circle indicates clustered location of microseismicity. The locking line lies in the transition zone between the two segments nearly corresponding with the MCT zone and is characterized by the clustering of microseismicity. Strain accumulates in the transition zone during interseismic period and released through earthquakes in the locked segment

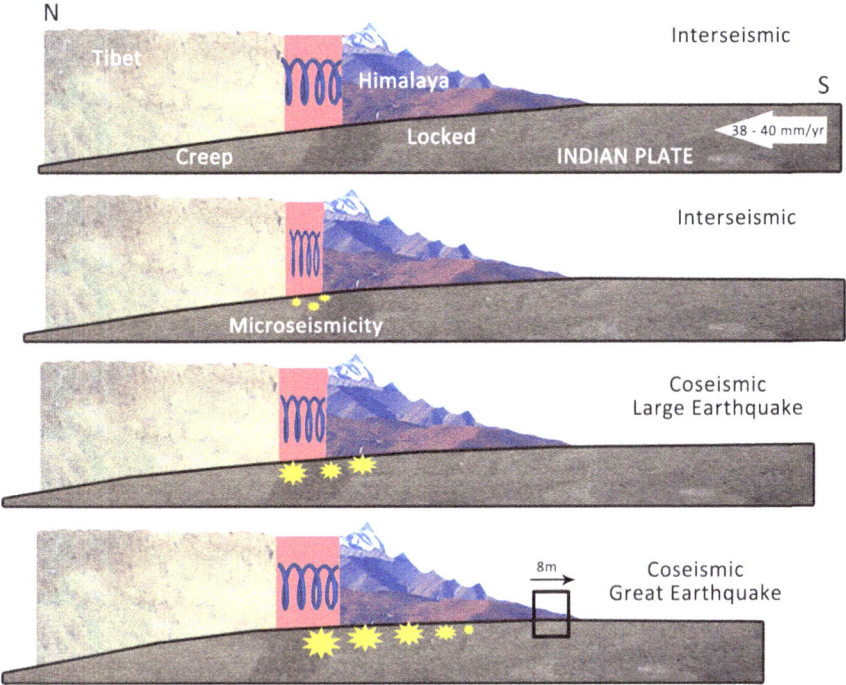

Fig. 1.8 A simplified cartoon showing mechanics of earthquake in Himalaya. Large earthquake produces emergent surface faulting as out-of-sequence thrusting (e.g., 2005 Kashmir earthquake), and sometime it remains blind (e.g., 2015 Gorkha earthquake). Great earthquake produces surface faulting at the toe of the active wedge where mountain meets the Ganga Plains. 8-m co-seismic slip is the maximum for an event estimated from the parametric study using scarp geometry of trenched fault zone across the HFT (Jayangondaperumal et al. 2013). Both MHT (1255 A.D. event) and wedge (Kashmir earthquake-2005) type-related earthquakes are possible

The geodetic observations imply that the strain accumulated in the locked segment is largely released through co-seismic deformation as a result of the great and large earthquakes. This interpretation is corroborated with the observation that the epicenter of the 1934 Bihar–Nepal earthquake of Mw 8.2 was located in the Lesser Himalaya of eastern Nepal, whereas the surface rupture of the same earthquake is recorded on the MFT at Sub-Himalayan front (Sapkota et al. 2013; Bollinger et al. 2014).

1.3 Regional Framework

Uttarakhand State constitutes the Garhwal and Kumaun regions forming a part of the northwest Himalaya. The geology of the Lesser Himalaya of Uttarakhand is described in detail by Valdiya (1980) (Fig. 1.9) and of High and Tethys Himalaya

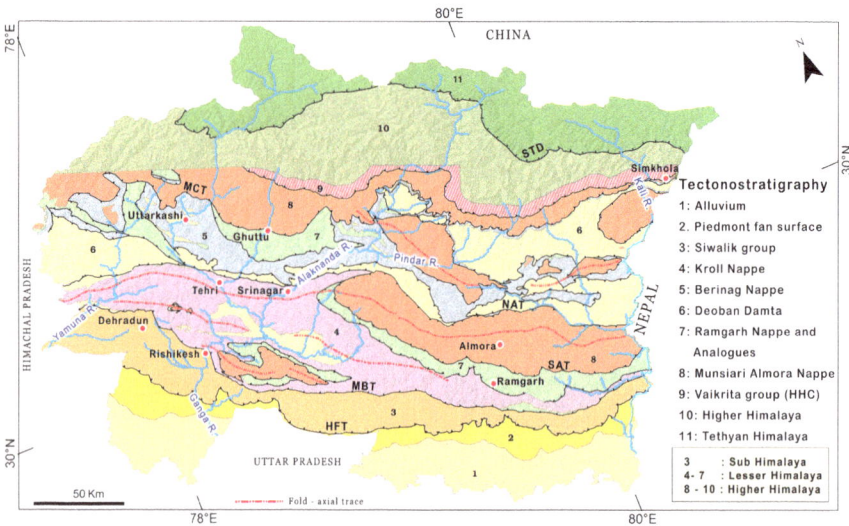

Fig. 1.9 Map showing the tectono-stratigraphic framework of Uttarakhand (modified after Valdiya 1980)

by Sinha (1981). The geology of Garhwal and Kumaun is also summarized in 'Geology of Western Himalaya' (Thakur 1992). The Uttarakhand is divided longitudinally into five physiographic–tectonic zones. The zones are: the Tarai-Bhabar zone, the Sub-Himalaya zone, the Lesser Himalaya zone, the High Himalaya zone, and the Tethys Himalaya zone.

1.4 Uttarakhand Tectonic Setting

1.4.1 Tarai-Bhabar Zone

The Tarai-Bhabar zone lies south of the Sub-Himalayan Siwalik Ranges and extends further south to the Gangetic Alluvial Plain between the Yamuna and the Sharda rivers. The Tarai-Bhabar is made up of coalescing alluvial fans formed by sediments brought from the hinterland and debouching over the piedmont zone. Boulder-to-pebble size clastics are deposited in the proximal part and sand, silt and mud toward the distal part. The piedmont zone developed near the mountain front is characterized by rolling topography. The rivers draining through the Tarai area are generally seasonal, remaining dry or with little runoff, but are flooded during the monsoon seasons. The major Himalayan rivers Yamuna, Ganga, Kosi, Ramganga, and Sarda drain through the Tarai region leaving behind the vast alluvial plain (Fig. 1.10). Two piedmont surfaces are recognized in the piedmont zone based on the soil characteristics and OSL ages. The Old Piedmont Zone (OPZ) ranges in

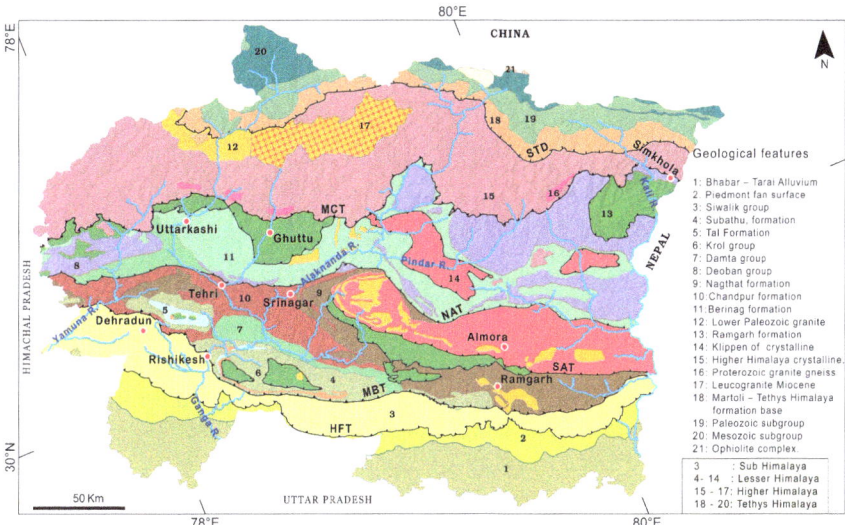

Fig. 1.10 Geological map of the Uttarakhand, Garhwal—Kumaun Himalaya (modified from the Thakur and Rawat map in Thakur 1992)

OSL age between 10 ka and 4 ka, and the Young Piedmont Zone (YPZ) is < 0.5 ka (Parkash et al. 2000; Thakur and Pandey 2004; Yeats and Thakur 2008). The piedmont zone to the south of Dehra Dun, between rivers Yamuna and Ganga, extends for ~ 15 km south of the frontal Siwalik range with a characteristic raised topography of the OPZ, which stands out from the surrounding YPZ and the floodplains as 5–15 m high isolated hillocks and small plateaus. The proximal parts of the OPZ are eroded away, and only the middle to distal part is observed, for example the NE–SW trending Biharigarh ridge.

1.4.2 Sub-Himalayan Zone

The Sub-Himalaya zone or the Siwalik Hills lie between the Himalayan Frontal Thrust (HFT) to the south and the Main Boundary Thrust (MBT) to the north. The boundary between the abruptly risen Siwalik range front and the Gangetic Alluvial Plain defines the HFT that abuts the Siwalik strata against the alluvium. The MBT demarcates a tectonic boundary between the Cenozoic Siwalik Group and the Proterozoic–lower Cambrian Krol Super Group of the Lesser Himalaya. In the Garhwal Sub-Himalaya, the Siwaliks, consisting of the Lower, Middle, and Upper Siwalik Formations, constitute a part of the foreland basin. In the Garhwal frontal Siwalik range, the Middle Siwalik sandstone is overlain by the Upper Siwalik conglomerate, both dipping NE 30°. Magnetic-polarity stratigraphy (MPS) of the Middle and Upper Siwalik strata in the adjoining region gives ages ranging from

6.6–4.8 Ma and 5–0.5 Ma, respectively (Sangode et al. 1996; Kumar et al. 2003). The Sub-Himalaya zone in the frontal part is characterized by the intermontane basins of Dehradun (Thakur and Pandey 2004; Jayangondaperumal et al. 2010a, b) and Kotadun (Goswami and Pant 2007) filled by post-Siwalik Dun gravels.

1.4.3 Lesser Himalayan Zone

The Lesser Himalayan zone lies between the Main Boundary Thrust (MBT) to the south and the Main Central Thrust (MCT) to the north. Two principal tectonic elements are recognized in the Lesser Himalaya Zone of the Kumaun–Garhwal Himalaya: the rock formations of the Lesser Himalaya basin and the klippen units of the crystalline thrust sheets, the latter overlying the former (Fig. 1.10). The Lesser Himalaya sequence consists of Chakrata and Rautgara Formations of Damtha Group at the base followed by the Deoban and Mandhali Formations of the Tejam Group, the Chandpur and Nagthat of the Jaunsar Group, the Blaini, Krol-Tal of the Mussoorie Group, and the Bansi and Subathu of the Sirmur Group (Valdiya 1980). These stratigraphic units are overlain by crystalline thrust sheets (called klippen) of Satengal, Lanesdowne, Almora, Baijnath, Askot, Chaukori, and Chiplakot (Jayangondaperumal and Dubey 2001). The klippen represent the eroded remanents of the Ramgarh and Munsiari thrust sheets, which are correlated with the thrust units underlying the Vaikrita Thrust (MCT–II) in the root zone. The linkage between the crystalline klippen in the lesser Himalaya and the crystalline thrust sheets below the MCT is validated through correlation of isotope geochemical signatures (Ahmad et al. 2000). In terms of structure, the Lesser Himalaya zone represents a duplex thrust system comprising horses of Beringag, Chakrata–Rautgara, and Deoban Formations overlain by the Ramgarh crystalline along the Ramgarh Thrust as roof thrust and the MHT as the floor thrust (Srivastava and Mitra 1994; Celerier et al. 2009). Based on field, structure, and paleomagnetic/anisotropy of magnetic susceptibility (AMS) data, an alternative model, negating large translation along the klippen detachment thrust from an assumed root zone in the Higher Himalaya, was also proposed (Jayangondaperumal 1998; Jayangondaperumal and Dubey 2001; Dubey and Jayangondaperumal 2005).

1.4.4 Higher Himalayan Zone

The High Himalayan zone constituting the main metamorphic belt of the Himalaya occupies the highest topographic line. It is 10–15 km-thick crystalline zone of metamorphics and granitoids. The metamorphic sequence shows progressive regional metamorphism ranging from biotite to sillimanite grade. The granitic gneisses and granites belonging to early Proterozoic, Cambrian, and Tertiary ages constitute an integral component of the metamorphics. The High Himalaya

Crystalline (HHC) lies between the Main Central Thrust (MCT) and the Tethys Himalaya sequence. The MCT, representing a shear zone, brings the HHC to override the Lesser Himalaya formations. According to Valdiya (1980), the HHC constitutes the Vaikrita Group underlain by the Munsiari and Ramgarh nappes. The base of the Vaikrita Group is demarcated by the Vaikrita Thrust, which is the sensu stricto MCT (Fig. 1.9). To the north, the HHC slab is overlain by the late Precambrian to Cretaceous Lower Eocene sequence of the Tethys Himalaya. The contact between the two tectonic zones is demarcated by the low-angle N-dipping normal fault called the South Tibetan Detachment (STD). The HHC is considered as representing the remobilized basement of the Indian crust incorporated in the Himalayan orogen (Hodges 2000; Yin 2000; Robinson et al. 2006), whereas in the channel flow model proposed in Nepal, the HHC was extruded from mid-crustal level of southern Tibet crust of low-velocity zone to the Himalayan orogen between the STD and the MCT driven by rapid erosion at the topographic front of the High Himalaya (Beaumont et al. 2001; Hodges et al. 2004).

1.4.5 Tethys Himalayan Zone

The Tethys Himalaya sequence is very well exposed in the drainage basins of the Goriganga, Dhauliganga, Darma, and Kali rivers in Kumaun and Niti and Malla Johar in Garhwal. Heim and Gansser (1939) gave the first systematic account of the biostratigraphy of the region. Later workers revised the stratigraphy of the Tethys Himalaya culminating into a special publication (Sinha 1989, and references therein). Late Precambrian Martoli Formation, equivalent to Upper Haimanta, constitutes the base of the Tethys sequence which is succeeded by fossiliferous Cambrian to Upper Cretaceous–Lower Eocene Formations. In Malla Johar area, the ophiolite complex klippe comprising ultramafic–serpentinite and volcanics overlie the Mesozoic sequence. They represent a remnant of the abducted oceanic crust slab from the Indus–Tsangpo Suture zone. The contact between the Martoli Formation and the underlying HHC was designated as the Trans-Himadari Fault by Valdiya (1980). This contact is now interpreted as the normal fault and referred on the regional scale as the South Tibetan Detachment (STD) (Fig. 1.10). The Tethys Himalaya sequence represents sediments deposited on the north-facing passive margin of the southern Tethys.

1.5 Crustal Structure

In the Garhwal Himalaya, the National Geophysical Research Institute (NGRI), Hyderabad, installed a network of 21 three-component broadband seismic stations along a SW–NE line in the Alaknanda valley during 2005–2006, covering across the area between the surface traces of the MBT and the STD (Fig. 1.11a).

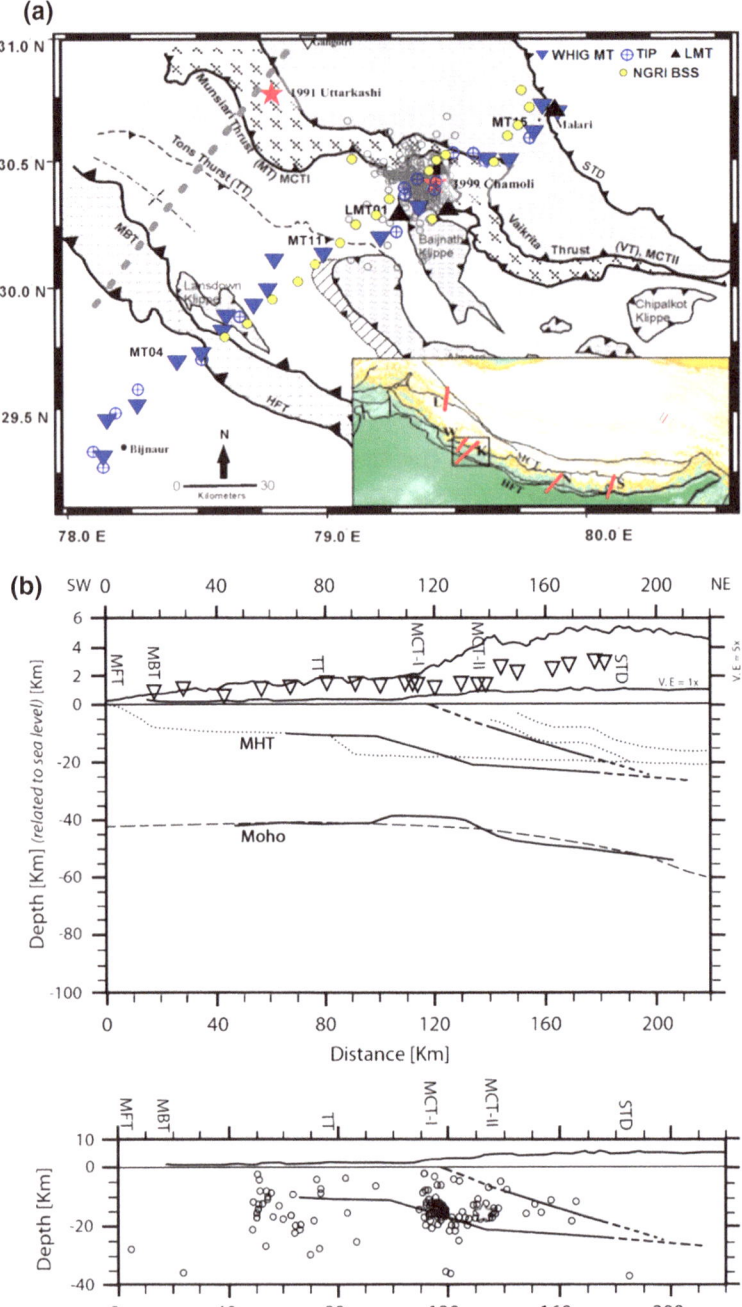

◄**Fig. 1.11** **a** Map showing the regional tectonic framework of the Garhwal–Kumaun Himalaya with locations of broadband seismometers (dots) and the magnetotelluric (MT) stations (inverted triangles) of WIHG along the NW–SE transect between the HFT and the STD. Locations of broadband seismic stations (BBS) on spatially coincident seismic survey are also marked (yellow dots). Stars depict the epicenters of 1991-Uttarakashi and 1999-Chamoli earthquakes. Open gray circles are aftershocks of 1999-Chamoli earthquake. The gray thick-dashed line is MT profile in the west Garhwal Himalaya (Israili et al. 2008). The locations of the MT profiles in NW Himalaya (L), passing through the epicenters of Uttarkashi (U) and Chamoli earthquakes (C) in Garhwal Himalaya, Nepal (N) and Sikkim (S) Himalaya are shown as bars in the inset. **b** *Top:* Using Ps receiver function analysis, crustal structure and Moho of the Garhwal Himalaya transect is imaged, shows Main Himalayan Thrust (MHT), detachment at the base of the Himalayan Thrust wedge with ramp and flat geometry (Rawat et al. 2014). *Bottom:* Schematic structural cross section across the Garhwal Himalaya showing the MHT with flat and ramp geometry and clustering of microseismicity in the ramp segment

Lithospheric structure of this segment of the Himalaya is deduced through the Ps receiver function analysis of 245 events with Mw $\geq$ 5.5 (Caldwell et al. 2013). In seismic images, the upper flat lies ~ 10 km below sea level beneath the Lesser Himalaya and dips $\sim 2°$. It connects to a mid-crustal ramp which extends to ~ 20 km depth with $\sim 16°$ dip. Lower flat is 20–25 km below sea level and dips $\sim 16°$. Moho beneath the Sub- and Lesser Himalayas is nearly horizontal and at the depth of 35–45 km (Fig. 1.11b). It deepens to 50 km or more under the High Himalaya. In structural cross-sectional interpretation, the ramp lies beneath the area between the Munsiari Thrust (MCT–I) and the Vaikrita Thrust (MCT–II) south of the High Himalaya topographic front (Fig. 1.11b). Magnetotelluric resistivity measurements along a profile between the Himalayan Frontal Thrust and the South Tibet Detachment indicate a low-angle northeast dipping intra-crustal high conducting layer (ICHCL) with a ramp at the transition from the Lesser Himalaya to the High Himalaya. The layer is within a depth range of 8–13 km. The high conductance ramp signifies low shear strength and high strain under the deviatoric stresses, which releases accentuated stresses into the brittle crust to generate small and more frequent earthquakes in the main seismicity belt of Himalaya (Rawat et al. 2014).

1.6   Seismicity

The microseismicity recorded instrumentally by the Wadia Institute of Himalayan Geology (WIHG) network shows clustering of events along the physiographic transition zone between the High and Lesser Himalaya and along the MCT zone (Paul et al. 2015) (Fig. 1.12). In our memory, Uttarakhand was struck by two moderate earthquakes, the 1991 Uttarkashi and the 1999 Chamoli. There is a historical account of 1803 earthquake causing severe damage and destruction largely in Garhwal and moderate-to-minor effect in the adjoining areas of Kumaun and

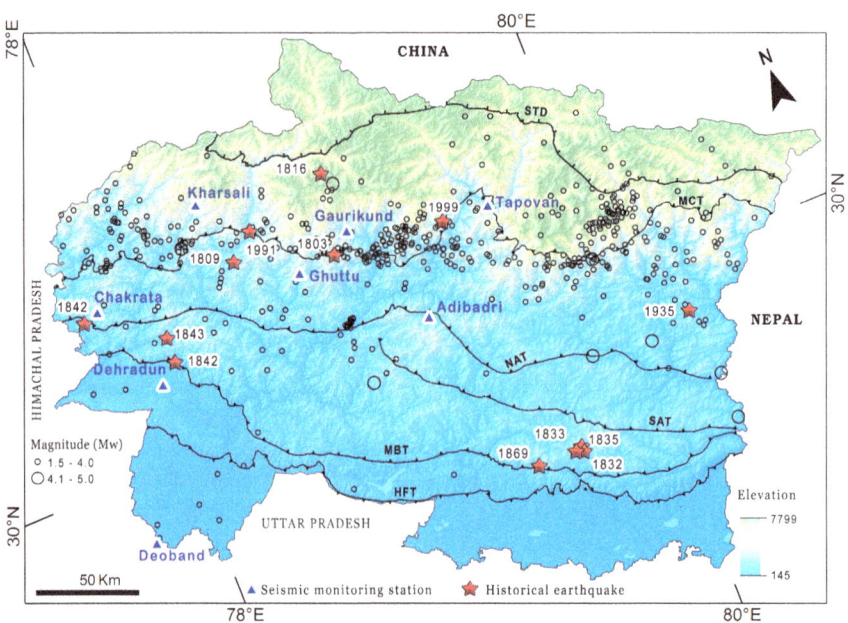

Fig. 1.12 Seismicity map of Uttarakhand, Garhwal–Kumaun Himalaya. The microseismicity plot is based on data generated through the WIHG network. The microseismicity is clustered in a belt trending NW–SE Himalayan regional trend and is distributed beneath the topographic front and south of the Greater Himalaya. Epicenters of moderate earthquakes derived from the historical archives are also shown in the map (Paul et al. 2015)

Sirmaur as well as in far-off places like Mathura, Aligarh, and Delhi. The details of these three earthquakes are described in the following section. There are several other earthquakes of magnitude Mw > 5 mentioned in the catalogue (Chandra 1992) compiled from the Indian Meteorological Department (Fig. 1.5b).

1.6.1   Uttarkashi Earthquake in 1991

The 1991 Uttarkashi earthquake occurred over a low-angle reverse fault south of the Vaikrita Thrust (MCT–II). About 700 people lost their lives, and considerable damage was inflicted to the property. Aftershocks and microseismicity monitored through an array (1979–1980 and 1984–1986) reveal the concentration around Bhatwari, north of Uttarkashi, at ~10-km depth (Note 1.1). The epicenter lies at the same depth underneath the MCT–II (Khattri et al. 1989; Thakur and Kumar 1994). The GSI network recorded three foreshocks (Mw $\geq$ 4.0) 48 h before and the two aftershocks Mw 4.2 and 5.2 within 24 h of the main shock. The total aftershocks recorded by the GSI network lie at a depth of 5–15 km.

Note 1.1

October 20, 1991 Uttarkashi earthquake

Earthquake parameters

Organization	Epicentral location	Focal depth (Km)	Body-wave magnitude Richter	Intensity Max.MMS
USGS	30.780°N	10.3	6.5	
	78.774°E			
IMD	30.750°N	12	6.6	VIII+
	78.860°E			

Fault plane solution (USGS, EDR bull.)

Plane/axis	Strike	Dip	Slip
Nodal plane 1	116°	85°	90°
Nodal plane 2	296°	05°	90°
	Plunge	Azimuth	
Pressure axis	40	26	
Tension axis	50	26	

Damage 658 persons killed, 5066 injured, 42,400 houses damaged

Based on aftershocks distribution, the Uttarkashi Fault, a bedrock-mapped fault was interpreted as the causative fault (Kayal 1996). In another monitoring experiment, 66 aftershocks (coda magnitude Mw ∼ 2.5–4.5) were recorded between October 1991 and March 1992 at Pala, near Bhatwari seismic station of WIHG. The composite fault plane solution derived from aftershocks indicates reverse faulting. The preferred fault plane is NP 2 indicating thrust movement along a NE-dipping shallow plane. Inversion of recorded accelerograms shows a complex rupture process, indicating 1.5-m slip maxima occurred 10 km west and 15 km southwest of the hypocenter along the detachment plane (Cotton et al. 1996).

1.6.2   Chamoli Earthquake in 1999

The 1999 Chamoli earthquake of magnitude ∼ 6.4 devastated the Chamoli and neighboring villages inflicting about 500 casualties (Note 1.2). The fault plane solution of Chamoli earthquake indicates reverse faulting with plane dipping NNE at low, 15–20° angle. The epicenter lies within the regional microseismicity belt located south of the Vaikrita Thrust in the MCT zone (Fig. 1.12). The main shock was followed by aftershocks, which continued for about three months. The aftershocks recorded by the WIHG network lie below the MCT at 7–17 km depth over the inferred ramp region (Thakur and Kumar 1994). The GSI network shows

aftershocks mostly clustered between 7 and 17 km depth and distributed south of
the MCT. In their model proposed, the earthquake occurred at the junction of the
Alaknanda Fault and the Basement Thrust at a depth of 21 km south of the MCT (Kayal
et al. 2003). In SAR data analysis of the Chamoli earthquake indicates rupture
dimensions of 13 ± 3 km along strike and 10 ± 3 km down-dip and surface deformation in a region of dimensions 30 km by 40 km with a maximum co-seismic uplift
of ~ 60 mm (Satyabala and Bilham 2006).

Note 1.2
March 29, 1999 Chamoli earthquake

Earthquake parameters				
Organization	Epicentral location	Focal depth (Km)	Body-wave magnitude Richter	Intensity Max.MMS
USGS	30.550°N	15.0	6.4	VIII
	79.424°E			
IMD	30.408°N	21.0	6.8	VIII
	79.416°E			

Damage ~ 64 people killed, 500 injured, 74,000 people were economically affected

1.6.3   Garhwal Earthquake in 1803

An earthquake on September 1, 1803 destroyed the royal capital of Garhwal at
Srinagar and caused extensive damage to life and property in the Bhagirathi and
Alaknanda Valleys and the adjoining areas. The earliest report on the damages
affected by the earthquake is found in the accounts given by Raper (1810), Hodgson
(1822) and Baird-Smith (1843). Captain Raper who visited Garhwal during 1807–
1808 covered most of the villages in the Bhagirathi and Alaknanda Valleys and
documented damage to buildings and temples. The maximum damage to buildings
was reported at Srinagar, Barhat (Uttarkashi), and Devprayag. The damage was also
reported farnorth Badrinath including the Badrinath temple and extending to Ojha
Ghur on the east bank of Yamuna River (Fig. 1.13). Many parts of the Indo-Gangetic
Plains and Himalayan foothills also experienced the wrath of the 1803 earthquake.
Ancient temple at Mayapur near Haridwar was destroyed. Piddington (1804) reported
the collapse of rubble masonry houses and the main mosque at Mathura and observed
ground fissures and bubbling of water (liquefaction).

   Similar damages were also reported from Aligarh, and the Aligarh Fort under the
siege of the advancing British troops fell due to the earthquake. In Delhi, the top
story cupola of the thirteenth century monument Qutab Minar fell down by the
1803 earthquake (Sharma 2001). Based on analysis of such reports and other

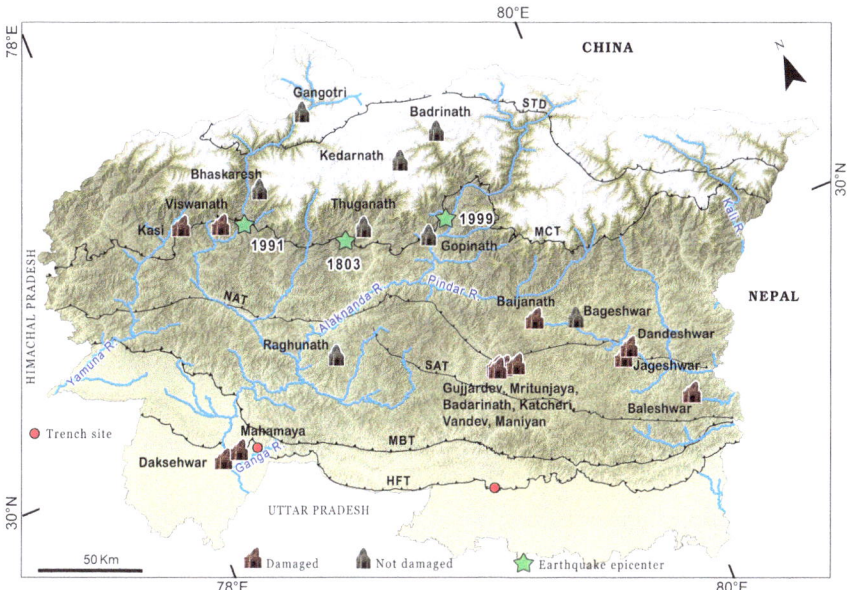

Fig. 1.13 Simplified tectonic map of the Garhwal and Kumaun regions of Uttarakhand compiled on ASTER DEM. Locations of historical heritage structures (temples) and epicenters of important earthquakes are shown. Epicenter locations are from Ambraseys and Douglas (2004). Red circles denote locations of the Ramnagar and Laldhang trench sites. Temples built in the A.D. 1100–1200 were damaged by an earthquake (A.D. 1344) prior to the A.D. 1803 Garhwal earthquake event (compiled and modified after Rajendran et al. 2013)

parameters, Rajendran and Rajendran (2005) prepared an isoseismal map (MSK scale), showing intensity IX–X over Srinagar and Devprayag, VII over Mathura and VI over Benaras. They assigned magnitude Mw 7.7 ± 0.4 to the 1803 event, close to magnitude Mw 7.5 given by Ambraseys and Douglas (2004). In addition to the above mentioned comparatively better documented earthquakes, Uttarakhand has experienced several other moderate earthquakes in the historical past. The events are reported in the IMD catalogue, given here in Note 1.3, and their locations shown in Fig. 1.10. The WIHG has been running a network of seismic stations for nearly a decade to monitor seismicity in the Uttarakhand region. The instrumentally recorded microseismicity is concentrated in a ∼50 km-wide belt south of the topographic front of the High Himalaya at a depth of 12–23 km covering the MCT zone between the Munsiari and Vaikrita Thrusts (Arora et al. 2012; Paul et al. 2015). This microseismicity belt is northwest continuation of the microseismicity distribution recorded in Nepal (Pandey et al. 1999; Jouanne et al. 1999). The microseismicity is clustered in the mid-crustal ramp segment of the MHT both in central Nepal and Garhwal (Pandey et al. 1999; Caldwell et al. 2013).

Note 1.3

Historical earthquake record

S.No.	Year	Latitude	Longitude	Location
1	1803	30.3	78.8	U.Ganga_Sirmoor & Garhwal
2	1809	30.7	78.5	Garhwal
3	1816	30.9	79	Gangotri_U.Valley of Ganga
4	1831	29.4	79.6	Lohaghat
5	1832	29.4	79.6	Lohaghat
6	1833	29.4	79.6	Lohaghat
7	1835	29.4	79.6	Lohaghat
8	1842	30.7	77.8	Mussoorie, Simla, NW provinces
9	1842	30.4	78.1	Mussoorie
10	1843	30.5	78.1	Landour
11	1869	29.4	79.4	Nainital
12	1935	29.75	80.25	Nepal_India border region
13	1991	30.45	78.46	Uttarkashi
14	1999	30.512	79.403	Chamoli

Source IMD Catalogue and Chandra (1992)

1.7   Geodetic (GPS) Studies

In northwest Himalaya, the GPS velocity field measurements across the rupture zone of the 1905 Kangra earthquake and the western portion of the adjoining region to the east reveal that the HFT is locked over a width of ~100 km where strain is building up to be ruptured during a future earthquake. The GPS measurements in the region indicate the best-fit slip rate for the currently locked HFT is 14 ± 1 mm (Banerjee and Burgmann 2002; Ponraj et al. 2011) (Fig. 1.14). Similarly, GPS network measurements between Dehradun reentrant and Harsil and Bhatwari in the MCT zone indicate that the ~100 km-wide segment between the Sub-Himalayan front and the southern front of High Himalaya is locked and a slip of 10–12 mm/year slip is consumed aseismically through creep under the High Himalaya where the Indian plate descends at greater depth of ~20 km (Banerjee and Burgmann 2002). This value of slip rate closely corresponds to the long-term late Holocene slip rate estimated through balanced cross section (Power et al. 1998) and using uplifted strath terraces as geomorphic markers (Wesnousky et al. 1999; Kumar et al. 2006).

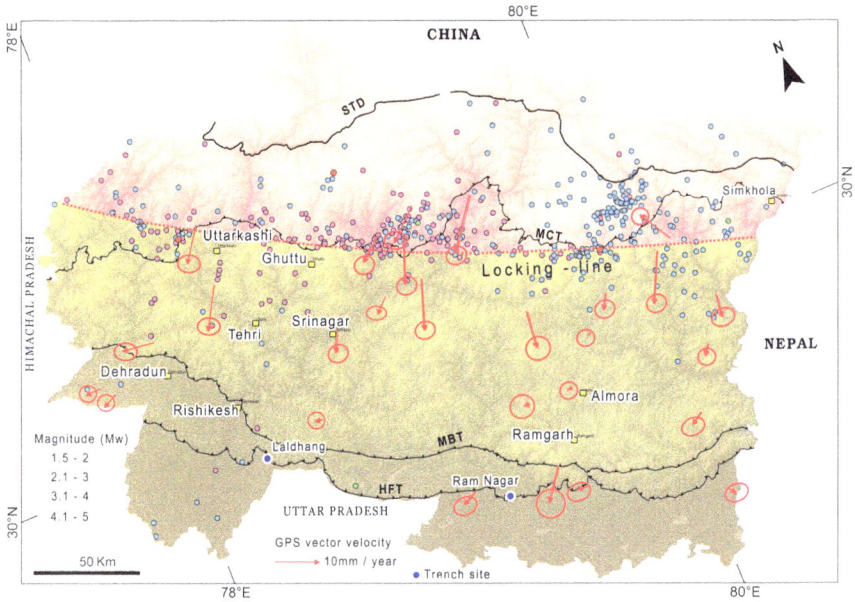

Fig. 1.14 Simplified map showing microseismicity and GPS vector velocity. Estimated site velocities at GPS stations in the Kumaun-Garhwal Himalaya with respect to the fixed Indian frame (modified after Banerjee and Burgmann 2002 and Ponraj et al. 2011). The locking line is a term to describe the transition from the fully locked part of the MHT to creeping Indian plate. In other words, the surface projection of interface between locked segment of MHT and creeping part lies on the north of Higher Himalaya is known as locking line. Position of locking line is variable along its strike of entire Himalayan arc

1.8 Paleoseismological and Archaeoseismological Investigations

Paleoseismological studies have been made by several workers along the Sub-Himalayan front in the Kumaun and Garhwal (Kumar et al. 2006; Rajendran et al. 2015; Malik et al. 2016; Jayangondaperumal et al. 2017b). Trenches were excavated normal to the NW–SE trending fault scarps along the HFT front in late Holocene sediments. Laldhang and Ramnagar trenches formed a part of the other three more trenches made at Rampur Ganda, Kala Amb, and Chandigarh. Radiocarbon ages of displaced sediments from the trenches indicate surface rupture took place between ∼A.D. 1278 and ∼A.D. 1433 (Kumar et al. 2006). Rajendran et al. (2015) excavated a new trench adjacent to the Ramnagar trench at Belparao of Kumar et al. (2006). They proposed two earthquake events that occurred during A.D. 1050–1259 and A.D. 1259–1433. These events were correlated with the A.D. 1255 and A.D. 1344 earthquakes reported in Nepal (Mugnier et al. 2013; Bollinger et al. 2016). About ∼8 km northwest of Belparao trench, near Dhol village, in another trench, Malik et al. (2016) reported three earthquake events. Of the three, the event–II occurred between A.D. 1294 and 1587, and the event–III occurred between A.D. 1750 and 1587. The event–II was correlated with the historical A.D. 1505 earthquake of Tibet–Nepal border and the event–III with the historical 1803 earthquake of Garhwal. The event–II period range A.D. 1294–1587 closely corresponds with the A.D. 1278–1433 event reported by Kumar et al. (2006).

More recently, modeling of published age data of five trenches (Kumar et al. 2006) using Bayesian statistical program OxCal indicates an event, coeval on the HFT, corresponds to a historically documented A.D. 1344 earthquake reported from Nepal (Pant 2002) with an estimated magnitude of >8.6 Mw (Jayangondaperumal et al. 2017a, b). Based on the damage pattern to the eleventh century temples at Garhwal and Kumaun, Rajendran et al. (2013) suggested that the rupture of the A.D. 1255 earthquake of Nepal, which killed king Abhay Malla and inflicted major damage to the Kathmandu Valley, extended westward to the Kumaun and Garhwal regions of Uttarakhand.

Geoarcheological study of an ancient archeological site at Khajnawar in the piedmont zone south of the Dehradun Valley reveals that the settlement with collapsed walls was most probably damaged by the 1803 earthquake (Thakur et al. 2010).

1.9   Reactivation of the Hinterland Faults

1.9.1   Reactivation of MCT

In the Bhagirathi river valley, Th-Pb ion microprobe ages of monazite dated in rock thin sections from the HHC are Eocene (38.0 ± 0.8 Ma) to Miocene (19.5 ± 0.3 Ma), which are consistent with the burial of the unit during imbrications of the northern Indian margin and subsequent exhumation due to reactivation of the MCT. Samples directly beneath the MCT yield Monazite ages of 4.3 ± 0.1 and 4.5 ± 1.1 Ma, and hydrothermal monazites record Th-Pb ages of 1.0 ± 0.5 Ma and 0.8 ± 0.2 Ma. This age data indicates out-of-sequence thrusting and reactivation consistent with critical-taper wedge model of the Himalaya (Catlos et al. 2007).

1.9.2   Reactivation of the Srinagar Thrust

The Srinagar Thrust in Garhwal is western continuation of the North Almora Thrust (NAT). In Srinagar area, the Srinagar Thrust, dipping ∼ SW at moderate to steep angle, separates the Pauri phyllite, earlier considered as Chandpur Formation of the Krol Group, from the underlying quartzites and volcanics of the Rautgarah (Damtha) Formation (Valdiya 1980). The Alaknanda River flows through a narrow gorge both north and south of the wide valley of Srinagar (Sati et al. 2007). The wide valley forms a small intermontane basin, 6 × 3 km, showing well-developed late Quaternary terraces over both sides of the Alaknanda River. Six levels of terraces are recognized and dated (Juyal et al. 2010; Ray and Srivastava 2010). Two major phases of aggradation separated by an accelerated uplift phase are recognized. The older aggradation event occurred around ~ 18 ka, and the younger phase lasted between 15 and 8 ka (Juyal et al. 2010). Ray and Srivastava (2010) inferred that aggradation occurred during 26–14 ka characterized through three phases of alluviation with cut-fill terraces separated by two periods of bedrock incision. The northern margin of the Srinagar basin is characterized by an abrupt rise of the mountain front from the relatively flat surface of the Chauras made up of fluvial terraces. The Srinagar Thrust, trending NW–SE, extends along the south-facing mountain front. The mountain front is characterized by well-developed triangular facets indicative of normal faulting, which is best observed from the opposite side of the mountain on the left flank of the Alaknanda River.

1.10 Himalayan Frontal Fault System

The Himalaya rises abruptly from the Indus-Ganga-Brahmaputra Alluvial Plain with a physiographic and tectonic boundary that is expressed in the form of a mountain front scarp. The mountain front scarp is either a fold scarp or a reverse fault scarp. A fault scarp is defined as tectonic landforms corresponding, or roughly coincident, with a fault plane that has displaced the ground surface. The trace of the boundary between the mountain front and the alluvial plain is defined as the Himalayan Frontal Thrust (HFT) or also designated as the Main Frontal Thrust (MFT). The thrust is emergent to the surface and can be clearly demarcated at many places, whereas at other places its expression is modified due to erosion, or the thrust may remain blind in case of a frontal anticline. The Himalayan Frontal Fault System encompasses the tectonic zone between the MBT and the HFT. The Himalayan Frontal Fault system includes the HFT and the overlying frontal anticlines and their complementary synforms (duns) (Yeats et al. 1992; Thakur et al. 2007; Thakur 2013; Yeats and Thakur 2008; Jayangondaperumal et al. 2010b). The HFF system has been studied in the area between the Rivers Ganga and Beas recently by several workers. In the Dehradun area of Garhwal, Mohand anticline was interpreted as fault-bend fold developed over the HFT during late Quaternary (Thakur et al. 2007), and 13-15 mm/year shortening was estimated across the HFT (Wesnousky et al. 1999). The paleoseismological investigations show that the HFT was reactivated by the surface ruptures of past historical earthquakes (Kumar et al. 2001, 2006).

1.10.1 Reactivation of MBT

The MBT was developed subsequent to the MCT in the foreland propagating thrust system. The timing of the MBT is constrained around late Miocene in northwest Himalaya (~ 10 Ma) (Meigs et al. 1995) and Pliocene in Nepal (~ 5 Ma) (DeCelles et al. 2001; Robinson and Pearson 2013). The MBT is largely inactive; however, its reactivation is reported in small segments in Nepal and Garhwal (Nakata 1972; Valdiya 1980; Mugnier et al. 1994). In the Kumaun Sub-Himalaya, reactivation of the MBT is expressed in active normal faulting along the MBT fault trace (Valdiya 1992a, b; Kothiari et al. 2010; Philip et al. 2017). In the Gaula Valley at Logar, a semi-circular fan is developed at the mountain front of the MBT. The fan deposit is displaced by a WNW trending normal fault characterized by a north-facing scarp 2.5 km in length and 37 m in height (Luirei et al. 2014; Philip et al. 2017). The active fault trace is coincident with the MBT trace, which is characterized with steeply north-dipping sandstone of the Logar Formation equivalent of Dharamsala/Siwalik Formation underlies the volcanic of the Krol Group, the former constituting the footwall and latter the hanging wall of the MBT. In Dehradun near Rajpur, crushed Chandpur Formation of Krol Group nappe overrides the Dun gravels along the MBT. The Chandpur consisting of phyllitic shale and siltstone overlies the unconsolidated Dun gravels showing crushing of phyllitic shale and formation of fault gauge (Aswathi 2012). The contact between the two dips NE $\sim 40°$. The dun gravels underlying the Chandpur on the footwall gives 30 ka OSL age. On the hanging wall Chandpur, another Dun gravel layer located ~ 50 m above the footwall gravel yields 20 ka OSL age. This observation at the locality on Mussoorie old track toll barrier indicates the reactivation of the MBT during the interval between 30 and 20 ka.

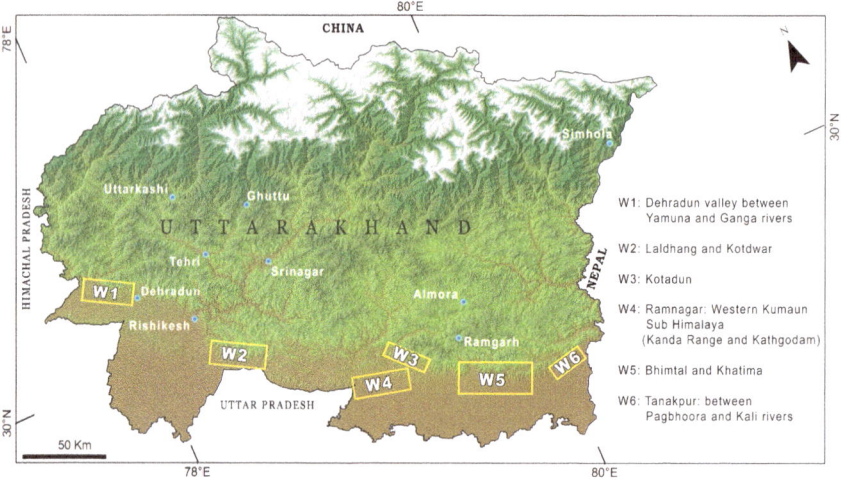

Fig. 1.15 Index map showing selected locations (Windows 1–6) where the active fault mapping has been undertaken and presented in the following chapters

1.10.2   Himalayan Frontal Thrust

In northwest Himalaya between the Beas and Sharda rivers, the HFT trending NW–SE is characterized by abrupt physiographic and tectonic breaks between the alluvial plain sediments and the Siwaliks of southern margin of the Sub-Himalaya front. The active fault scarps have been mapped along the HFT at several places between the Beas and Satluj rivers (Yeats and Thakur 2008; Malik et al. 2010a, b; Kumahara and Jayangondaperumal 2013; Jayangondaperumal et al. 2017a). South of Dehradun, between the Yamuna River and Mohand at the Sub-Himalayan front (Fig. 1.7), several strath terraces, 10–30 m high, from the stream grade in middle Siwalik were mapped and the terrace deposit was dated (Wesnousky et al. 1999). The estimated late Holocene shortening and fault slip rates across the HFT are 11.9 + 3.1 and 13.8 + 3.6 mm/year, respectively.

1.11   Mapping of Active Faults in Selected Windows

The Himalaya Mountains have grown from north to south. The principal faults, the MCT, the MBT, and the HFT (MFT), were developed progressively to the south; the MCT in middle Miocene, the MBT in late Miocene–Pliocene, and the HFT in Quaternary–Holocene. We have carried out our active fault mapping in the Sub-Himalaya zone of Uttarakhand, concentrating on the Sub-Himalayan front and the dun region. Based on reconnaissance field survey and consulting satellite imageries, active fault mapping (i.e., fault scarp) is directed to the selected areas referred here as the window (W) (Fig. 1.15). The selected windows are: W1—Dehradun valley between Yamuna and Ganga Rivers, W2—Laldhang and Kotdwar, W3—Kotadun, W4—Ramnagar: Western Kumaun Sub-Himalaya (Kanda Range and Kathgodam), W5—Bhimtal and Khatima, and W6—Tanakpur between Pagbhoora and Kali Rivers.

References

Ader T, Avouac JP, Liu-Zeng J, Lyon-Caen H et al (2012) Convergence rate across the Nepal Himalaya and interseismic coupling on the Main Himalayan Thrust: implications for seismic hazard. J Geophys Res 117:B04403. https://doi.org/10.1029/2011JB009071

Ahmad T, Harris N, Bickle M, Chapman H, Bunbury J, Prince C (2000) Isotopic constraints on the structural relationships between the Lesser Himalayan Series and the High Himalayan Crystalline Series, Garhwal Himalaya. Geol Soc Amer Bull 112:467–477

Ambraseys N, Bilham R (2000) A note on the Kangra Ms = 7.8 earthquake of 4 April 1905. Curr Sci 79:101–106

Ambraseys N, Douglas J (2004) Magnitude calibration of North Indian earthquakes. J Geophys Int 159:165–206

Armbruster et al (1978) The northwestern termination of the Himalayan Mountain Front: Active
    tectonics from microearthquakes January 1978. J Geophys Res Atmos 83(B1):269–282.
    https://doi.org/10.1029/JB083iB01p00269

Armijo R, Tapponnier P, Mercier J, Han T (1986) Quaternary extension in southern Tibet: field
    observations and tectonic implications. J Geophys Res 91(13):803–872

Arora BR, Gahalaut VK, Kumar N (2012) Structural control on along strike variation in the
    seismicity of northwest Himalaya. J Asian Earth Sci 57:15–24

Aswathi NS (2012) Project report on Neotectonic activity on Main Boundary Thrust near
    Dehradun, Garhwal Himalaya. A report submitted to Indian Academy of Science, Bangalore,
    p 30

Avouac J-P, Tapponnier P (1993) Kinematic model of active deformation in central Asia. Geophys
    Res Lett 20:895–898

Avouac J-P, Ayoub F, Lefrince S, Konca K, Hellembergner DV (2006) The 2005, Mw 7.6,
    Kashmir earthquake: sub-pixel correlation of ASTER images and seismic wave form analysis.
    Earth Planet Sci Lett 249:514–528

Avouac J-P, Meng L, Wei S, Wang T, Ampuero J-P (2015) Lower edge of locked Main
    Himalayan Thrust unzipped by the 2015 Gorkha earthquake. Nat Geosci 8:708–711

Baird-Smith R (1843) Memoir on Indian earthquakes. J Asia Soc Bengal 2(12):1029–1056

Banerjee P, Burgmann R (2002) Convergence across the northwest Himalaya from GPS
    measurements. Geophys Res Lett 29:30-1–30-4

Beaumont C, Jamieson RA, Nguyen MH (2001) Himalayan tectonics explained by extrusion of a
    low velocity crustal channel coupled to focused surface denudation. Nature 414:738–742

Bettinelli P, Avouac JP, Flouzat M, Jouanne F, Bollinger L, Willis P, Chitrakarm G (2006) Plate
    motion of India and Interseismic strain in the Nepal Himalaya from GPS and DORIS
    measurements. J Geod 80:567–589

Bilham R (1995) Location and magnitude of the 1833 Nepal earthquake and its relation to the
    rupture zones of contiguous great Himalayan earthquakes. Curr Sci 69:101–128

Bilham R, Larson K, Freymuller J (1997) Indo-Asian convergence rates in Nepal Himalaya.
    Nature 386:61–66

Bilham R, Gaur VK, Molnar P (2001) Himalayan seismic hazard. Science 293:1442–1444

Bollinger L, Sapkota SN, Tapponnier P, Klinger Y, Rizza M, Van der Woerd J, Tiwari DR,
    Pandey R, Bitri A, Bes de Berc S (2014) Estimating the return times of great Himalayan
    earthquakes in eastern Nepal: evidence from the Patu and Bardibas strand of the Main Frontal
    Thrust. J Geophys Res Solid Earth 119:7123–7163. https://doi.org/10.1002/2014jb01090

Bollinger L, Tapponnier P, Sapkota SN, Klinger Y (2016) Slip deficit in central Nepal: omen for a
    repeat of the 1344 AD earthquake? Earth Planets Space 68:12. https://doi.org/10.1186/s40623-
    016-0389-1

Bombolakis EG (1986) Thrust-fault mechanics and origin of a frontal ramp. J Struct Geol 8(3–
    4):281–290

Bombolakis G (1994) Applicability of critical-wedge theories to foreland belts. Geology 22:535–
    538

Brown LD, Zhao WJ, Nelson DK, Hauck M, Alsdorf D, Ross A, Cogan M, Clark M, Che XW
    (1996) Bright spots, structure, and magmatism in southern Tibet from INDEPTH seismic
    reflection profiling. Science 274:1688–1690

Burges WP, Yinm A, Dubey CS, Shen ZK, Kelty TK (2012) Holocene shortening across the Main
    Frontal Thrust zone in the eastern Himalaya. Earth Planet Sci Lett 357–358:152–167

Caldwell WB, Klemperer S, Lawrence JF, Rai SS, Ashish (2013) Characterizing the Main
    Himalayan Thrust in the Garhwal Himalaya, India with receiver function CCP stacking. Earth
    Planet Sci Lett 367:15–27

Catlos EJ, Dubey CS, Marston RA, Harrison TM (2007) Geochronologic constraints across the
    Main Central Thrust shear zone, Bhagirathi River (NW India): Implications for Himalayan
    tectonics. In Cloos M, Carlson WD, Gilbert MC, Liou JG, Sorensen SS (eds) Convergent
    Margin Terranes and Associated Regions: A Tribute to W.G. Ernst: Geological Society of
    America Special Paper 419, 135–151. https://doi.org/10.1130/2006.2419(07)

Celerier J, Harrison TM, Webb AAG, Yin A (2009) The Kumaun and Garwhal Lesser Himalaya, India: part 1. Structure and stratigraphy. Geol Soc Am Bull 121:1262–1280

Champel B, Van der BP, Mugnier JL, Leturmy P (2002) Growth and lateral propagation of fault-related folds in the Siwaliks of western Nepal: rates, mechanisms, and geomorphic signature. J Geophys Res 107(B6):2.1–2.18.

Chandra U (1992) Seismotectonics of Himalaya. Curr Sci 62:40–72

Chen S, Molnar P (1990) Source parameters of earthquakes beneath the Shillong plateau and the Northern Indo-Burman ranges. J Geophys Res 95:12527–12552

Cotton F, Camplillo M, Deschamps A, Rastogi BK (1996) Rupture history and seismotectonics of the 1991 Uttarkashi, Himalayan earthquake. Tectonophysics 258:35–51

De Bremaecker JC (1987) Is the oceanic lithosphere elastic or viscous? J Geophys Res 82:2001–2004

DeCelles PG, Gehrels GE, Quade J, Ojha TP (1998) Eocene–early Miocene foreland basin development and the history of Himalayan thrusting, western and central Nepal. Tectonics 17:741–765. https://doi.org/10.1029/98TC02598

DeCelles PG, Robinson DM, Quade J, Ojha TP, Carmala N, Garzione P, Copeland P, Upreti BN (2001) Stratigraphy, structure, and tectonic evolution of theHimalayan fold-thrust belt in western Nepal. Tectonics. 20:487–509

Delcaillau B, Carozza JM, Laville E (2006) Recent fold growth and drainage development: the Janauri and Chandigarh anticlines in the Siwalik foothills, northwest India. Geomorphology 76:241–256

Dubey AK, Jayangondaperumal (2005) One Pop-Up Klippen in the Mussoorie Syncline, Lesser Himalaya: Evidence from field and model deformation Studies. In Saklani PS (ed) Himalaya (Geological Aspects), vol 3, Satish Serial Publishing House, New Delhi, India, pp 203–222. ISBN: 9788189304041

Dunn JA, Auden JB, Ghosh AMN, Roy SC (1939) The Bihar-Nepal earthquake of 1934. Mem Geol Survey India 73:391 (reprinted 1981)

Feldl N, Bilham R (2006) Great Himalayan earthquakes and the Tibetan plateau. Nature 444:165–170. https://doi.org/10.1038/nature05199

Fielding EJ (2000) Morphotectonic evolution of the Himalayas and Tibetan plateau. In Summerfield MA (ed) Geomorphology and global tectonics. Wiley, Chichester, pp 201–222

Gahalaut VK, Chander R (1999) Geodetic evidence for accumulation of earthquake generating strains in the NW Himalaya near 75.5E longitude. Bull Seism Soc America 89:837–843

Gahalaut VK, Kundu B (2011) Possible influence of subducting ridges on the Himalayan arc and on the ruptures of great and major Himalayan earthquakes. Gondwana Res 21:1080–1088. https://doi.org/10.1016/j.gr.2011.07.021

Gavillot Y, Meigs A, Yule D, Heermance R, Rittenour T, Madugo C, Malik M (2016) Shortening rate and Holocene surface rupture on the Riasi fault system in the Kashmir Himalaya: active thrusting within the Northwest Himalayan orogenic wedge. Geol Soc Am Bull B31281-1

Goswami PK, Pant CC (2007) Geomorphology and tectonics of Kota–Pawalgarh Duns, Central Kumaun Sub-Himalaya. Curr Sci 92(5):685–690

Grujic D, Coutand I, Bookhagen B, Bonnet S, Blythe A, Duncan C (2006) Climatic forcing of erosion, landscape, and tectonics in the Bhutan Himalayas. Geology 34:801–804. https://doi.org/10.1130/G22648.1

Gualandi A et al (2016) Pre- and post-seismic deformation related to the 2015, Mw7.8 Gorkha earthquake, Nepal. Tectonophysics. https://doi.org/10.1016/j.tecto.2016.06.014

Hauck ML, Nelson KD, Brown LD, Zhao W, Ross AR (1998) Crustal structure of the Himalayan orogen at ∼ 90° east longitude from Project INDEPTH deep reflection profiles. Tectonics 17(4): 481–500

Heim A, Gansser A (1939) Central Himalaya: geological observations of the Swiss expedition, 1936. Mem Swiss Soc Nat Sci 73:245

Hodges KV (2000) Tectonics of the Himalaya and southern Tibet from two perspectives. Geol Soc Amer Bull 112:324–350

Hodges KV, Wobus C, Ruhl K, Schildgen T, Whipple K (2004) Quaternary deformation, river steepening, and heavy precipitation at the front of the Higher Himalayan ranges. Earth Planet Sci Lett 220:379–389

Hodgson JA (1822) Journey of a survey to the heads of the rivers, Ganges and Jumna. Asiatic Res 14:60–152

Hubbard MS, Harrison TM (1989) 40Ar/39Ar age constraints on deformation and metamorphism in the MCT zone and Tibetan slab, eastern Nepal Himalaya. Tectonics 8:865–880

Hussain A, Yeats RS, MonaLisa (2009) Geological setting of the 8 October 2005 Kashmir earthquake. J Seism 13(3):315–325

Israili M, Tyagi S, Gupta P, Niwas S (2008) Magnetotelluric investigations for imaging electrical structure of Garhwal Himalayan corridor, Uttarakhand, India. J Earth Syst Sci 117:189–200. https://doi.org/10.1007/s12040-008-0023-0

Jade S, Mukul M, Bhattacharyya AK, Vijayan MSM, Saigeetha J, Kumar Ashok, Tiwari RP, Kumar Arun, Kalita S, Sahu SC, Krishna AP, Gupta SS, Murthy MVRL, Gaur VK (2007) Estimates of inter seismic deformation in Northeast India from GPS measurements. Earth Planet Sci Lett 263:221–234

Jade S et al (2011) GPS-derived deformation rates in northwestern Himalaya and Ladakh. Int J Earth Sci (Geol Rund) 100:1293–1301. http://dx.doi.org/10.1007/s00531-010-0532-3

Jayangondaperumal (1998) Structural evolution of Mussoorrie Syncline, Lesser Himalaya, U.P., Unpublished Ph.D., Thesis, H.N.B. Garhwal University, Srinagar (Garhwal)

Jayangondaperumal R, Dubey AK (2001) Superposed folding of a blind thrust and formation of Klippen: results of anisotropic magnetic susceptibility studies from the Lesser Himalaya. J Asian Earth Sci 19:713–725

Jayangondaperumal R, Thakur VC (2008) Kinematics of Coseismic Secondary Surface Fractures on Southeastward Extension of the Rupture Zone of Kashmir Earthquake. Tectonophysics 446:61–76. https://doi.org/10.1016/j.tecto.2007.10.006

Jayangondaperumal R, Dubey AK, Kumar S, Wesnousky SG, Sangode SJ (2010a) Magnetic fabrics indicating Late Quaternary seismicity in the Himalayan foothills. Int J Earth Sci 99 (Suppl 1):265–278. https://doi.org/10.1007/s00531-009-0494-5

Jayangondaperumal R, Dubey AK, Sen K (2010b) Structural and magnetic fabric studies of recess structures in the western Himalaya: Implications for 1905 Kangra earthquake. In the Special issue on "Structural Geology-Classical to Modern Concept" edited by M.A. Mamtani. J Geol Soc India 75:212–225

Jayangondaperumal R, Wesnousky SG, Chaudhary B (2011) Near surface expression of Early to Late Holocene displacement along the Northeastern Himalayan Frontal Thrust at Marbang Korong Creek, Arunachal Pradesh, India. Bull Seism Soc America 101, 3060–3064. https://doi.org/10.1785/0120110051

Jayangondaperumal R, Mugnier JL, Dubey AK (2013) Earthquake slip estimation from the scarp geometry of Himalayan Frontal Thrust, western Himalaya: implications for seismic hazard assessment. Int J Earth Sci 102:1937–1955. https://doi.org/10.1007/s00531-013-0888-2

Jayangondaperumal R, Kumahara Y, Thakur VC, Kumar A, Srivastava P, Shubhanshu D, Joevivek V, Dubey AK (2017a) Great earthquake surface ruptures along backthrust of the Janauri anticline, NW Himalaya. J Asian Earth Sci 133:89–101. https://doi.org/10.1016/j.jseaes.2016.05.006

Jayangondaperumal R, Daniels RL, Niemi TM (2017b) A paleoseismic age model for large-magnitude earthquakes on fault segments of the Himalayan Frontal Thrust in the Central Seismic Gap of northern India. Quatern Int. https://doi.org/10.1016/j.quaint.2017.04.008

Jouanne F, Mugnier JL, PandeyMR Gamond J F, Le Font P, Serrurier L, Vigny C, Avouac PJ (1999) Oblique convergence in the Himalayas of western Nepal deduced from preliminary results of GPS measurements. Geophys Res Lett 26:1933–1936

Juyal N, Sundriyal YP, Rana N, Chaudhary, Shipra Singhvi AK (2010) Late quaternary fluvial aggradation and incision in the monsoon dominated Alaknanda valley, Central Himalaya, Uttarakhand, India. J Quat Sci 25:1293–1304

Kaneda H, Nakata T, Tsutsumi H, Kondo S, Sugito N, Awata Y, Akhtar S, Majid A, Khatak W, Awan A, Yeats RS, Hussain A, Ashraj M, Wesnousky SG, Kausar B (2008) Surface rupture of the 2005 Kashmir, Pakistan earthquake and its active tectonic implication. Bull Seism Soc Am 98:512–557

Kayal JR (1996) Earthquake source processes in northeast India: a review. Him Geol 17:53–69

Kayal JR, Sagina R, Singh OP, Chakraborty PK, Karunakar G (2003) Aftershocks of the March, 1999 Chamoli Earthquake and Seismotectonic structure of the Garhwal Himalaya. Bull Seism Soc Am 93(1):109–117

Keller EA, Pinter N (1999) Active tectonics: earthquakes, uplift, and landscape. Prentice-Hall, New Jersey, p 338

Khattri KN, Chander R, Gaur VK, Sarkar I, Kumar S (1989) New seismological results on the tectonics of the Garhwal Himalaya. Proc Indian Acad Sci Earth Planet Sci 98:91–109

Kondo H, Nakata T, Akhtar SS, Wesnousky SG, Sugito N, Kaneda H, Tsutsumi H, Khan AM, Khattak W, Kausar AB (2008) Long recurrence interval of faulting beyond the 2005 Kashmir earthquake around the northwestern margin of the Indo-Asian collision zone. Geology 36:731–734

Kothiari GC, Pant PD, Joshi M, Luirei K, Malik JN (2010) Active faulting and deformation of quaternary landform Sub Himalaya, India. Geochronometrica 37:63–71. https://doi.org/10.2478/v10003-010-0015-3

Krishnaswamy VS, Jalote SP, Shome SK (1970) Recent crustal movements in NW Himalaya and Gangetic plain. In: Proceedings of the 4th symposium earthquake engineering, Roorkee University, pp 419–439

Kumahara Y, Jayangondaperumal R (2013) Paleoseismic evidence of a surface rupture along the northwestern Himalayan Frontal Thrust (HFT). Geomorphology 180–181:47–56

Kumar S, Wesnousky SG, Rockwell TK, Ragona D, Thakur VC, Seitz GG (2001) Earthquake recurrence and rupture dynamics of Himalayan frontal thrust, India. Science 294:2328–2331

Kumar R, Ghosh SK, Sangode SJ (2003) Mio-Pliocene sedimentation history in the northwestern part of the Himalayan foreland basin, India. Curr Sci 84:1006–1113

Kumar S, Wesnousky SG, Rockwall TK, Briggs RW, Thakur VC, Jayangondaperumal R (2006) Paleoseismic evidence of great surface rupture earthquake along the Indian Himalaya. J Geophys Res 111:B03304. https://doi.org/10.1029/2004JB00309

Kumar S, Wesnousky SG, Jayangondaperumal R, Nakata T, Kumahara Y, Singh V (2010) Paleoseismological evidence of surface faulting along the northeastern Himalayan front, India: timing, size, and spatial extent of great earthquakes. J Geophys Res 115(B12422):20. https://doi.org/10.1029/2009JB006789

Lave J, Avouac JP (2000) Active folding of fluvial terraces across the Siwalik Hills, Himalayas of Central Nepal. J Geophys Res 105:5735–5770. https://doi.org/10.1029/1999JB900292

Lave J, Yule D, Sapkota S, Basant K, Madden C, Attal M, Pandey MR (2005) Evidence of a great Medieval earthquake (∼1100AD) in the Central Nepal. Science 307:1302–1305

LeFort P (1975) Himalayas: the collided range. Present knowledge of the continental arc. Am J Sci 275-A: 1–44

Luirei K, Bhakuni SS, Suresh N, Kothyari GC, Pant P (2014) Tectonic geomorphology and morphometry of the frontal part of Kumaun Sub-Himalaya: appraisal of tectonic activity. Z Geomorphol. https://doi.org/10.1127/0372-8854/2014/0134

Lyon-Caen H, Molnar P (1985) Gravity anamolies, flexure of Indian plate and the structure, support, and evolution of the Himalayan Ganga basin. Tectonics 4:513–538

Malik JN, Mathew G (2005) Evidence of paleoearthquakes from trench investigations across Pinjore Garden fault in Pinjore Dun, NW Himalaya. J Earth Syst Sci 114(4):387–400

Malik JN, Nakata T, Philip G, Suresh N, Virdi NS (2008) Active fault and paleoseismic investigation: evidence of historic earthquake along Chandigarh fault in the frontal Himalayan zone, NW India. Him Geol 29:109–117

I   Introduction

Malik JN, Sahoo AK, Shah AA, Shinde DP, Juyal N, Singhvi AK (2010a) Paleoseismic evidence from trench investigation along Hajipur fault, Himalayan Frontal Thrust, NW Himalaya. Implication of faulting pattern on landscape evolution and seismic hazard. J Struct Geol 32:350–361

Malik JN, Shah AA, Sahoo AK, Puhan B, Banerjee C, Shinde DP, Juyal N, Singhvi AK, Rath S (2010b) Active fault, fault growth and segment linkage along the Janauri anticline (fontal foreland fold), NW Himalaya, India. Tectonophysics 483:327–343

Malik JN, Naik SP, Sahoo S, Okumura K, Mohanty A (2016) Paleoseismic evidence of the CE 1505 (?) and CE 1803 earthquakes from the foothill zone of the Kumaon Himalaya along the Himalayan Frontal Thrust (HFT), India. Tectonophysics. http://dx.doi.org/10.1016/j.tecto.2016.07.026

Meigs AJ, Burbank DW, Beck RA (1995) Middle-late Miocene [>10 Ma] formation of the Main Boundary Thrust in the Western Himalaya. Geology 23:423–426

Middlemiss CS (1910) Kangra earthquake of 4th April, 1905. Mem Geol Surv India 39:1–409

Mishra RL, Singh I, Pandey A, Rao PS, Sahoo HK, Jayangondaperumal R (2016) Paleoseismic evidence of a giant medieval earthquake in the eastern Himalaya. Geophys Res Lett 43:5707–5715. https://doi.org/10.1002/2016GL068739

Molnar P (1984) Structure and tectonics of the Himalaya: constraints and implications of geophysical data. Ann Rev Earth Planet Sci 12:489–518

Molnar P (1986) The structure of mountain ranges. Sci Am, 70–79

Mugnier JL, Huyghe P, Chalaron E, Mascle G (1994) Recent movements along the Main Boundary Thrust of the Himalayas: normal faulting in an over-critical thrust wedge? Tectono physics, 238:199–2015

Mugnier JL, Leturmy P, Mascle G, Huyghe P, Chalaron E, Vidaln G, Husson L, Delcaillau B (1999) The Siwaliks of Western Nepal: I. geometry and kinematics. J Asian Earth Sci 17 (5-6):629–642

Mugnier JL, Huyghe P, Leturmy P, Jouanne F (2004) Episodicity and rates of thrust sheet motion in Himalaya (Western Nepal). In McClay (ed) Thrust Tectonics and Hydrocarbon Systems. pp 91–114. A. A. P. G. Mem, 82

Mugnier JL, Huyghe P, Gajurel AP, Becel D (2005) Frontal and piggy-back seismic ruptures in the external thrust belt of Western Nepal. J Asian Earth Sci 25:707–717

Mugnier JL, Gajure A, Huyghe P, Jayangndaperumal R, Jouanne F, Upreti BN (2013) Structural interpretation of the great earthquakes of the last millennium in the central Himalaya. Earth Sci Rev 127:30–47

Mullick M, Federica RRF, Mukhopadhyay D (2009) Estimates of motion and strain rates across active faults in the frontal part of eastern Himalayas in North Bengal from GPS measurements. Terra Nova 21:410–415

Nabelek J, HI-ClIMB Team (2009) Underplating in the Himalaya-Tibet collision zone revealed by the Hi-CLIMB experiment. Science 325(5946):1371–1374. https://doi.org/10.1126/1167719

Nakata T (1972) Geomorphic history and crustal movements of the foothills of the Himalayas. Sci Rep Tohoku Univ 22(7):39–177

Nakata T (1975) On quaternary tectonics around the Himalayas. Tohoku Univ Sci Rep 25:111–118. 7th Ser. (Geography)

Nakata T, Kumahara Y (2002) Active faulting across the Himalaya and its significance in the collision tectonics. Active Fault Res 2002(22):7–16

Nakata T et al (1998) First successful paleoseismic trench study on active faults in the Himalaya. Eos Trans AGU 79(45), Fall Meet Suppl., Abstract S22A-18

Naresh K, Sharma J, Arora BR, Mukhopadhyay S (2009) Seismotectonic Model of the Kangra–Chamba Sector of Northwest Himalaya: Constraints from Joint Hypocenter Determination and Focal Mechanism. B Seismol Soc Am 99(1):95–109. https://doi.org/10.1785/0120080220. February 2009.

Nawani PC, Khan SA, Singh AK (1982) Geologic cum geomorphic evolution of the western part of Chenab basin with special reference to Quaternary tectonics. Him Geol 12:264–279

Nelson KD et al (1996) Partially molten middle crust beneath Southern Tibet: synthesis of project INDEPTH results. Science 274:1684–1696

Ni J, Barazangi M (1984) Seismotectonics of the Himalayan Collision zone-geometry of the under thrusting Indian plate beneath the Himalaya. J Geophys Res 89:1147–1163

Pandey MR, Tandukar RP, Avouac JP, Lavé J, Massot, JP, (1995) Evidence for recentinterseismic strain accumulation on a mid-crustal ramp in the Central Himalaya of Nepal. Geophys Res Lett 22:751–758

Pandey MR, Tandulkar RP, Avouac JP, Vergne J, Heritier T (1999) Seismotectonics of the Nepal Himalaya from a local seismic network. J Asian Earth Sci 17:703–712

Pant MR (2002) A step toward a historical seismicity of Nepal. Adarsa 2:29–60

Parameswaran RM, Natrarajan T, Rajendran K, Rajendran CP, Mallick R, Wood M, Lekhak HC (2015) Seismotectonic of the April–May 2015 Nepal earthquakes: an assessment based on the aftershock patterns, surface effects and deformational characteristics. J Asian Earth Sci 111(1): 161–174

Parkash B, Kumar S, Rao MS, Giri SC, Kumar CS, Gupta S, Srivastava P (2000) Holocene tectonic movements and stress field in the western Gangetic plains. Curr Sci 79:438–449

Paul A, Prasath AR, Singh R (2015) Slip heterogeneities evaluated for earthquakes M > 4.0 using waveform modelling in the Garhwal region of Central Seismic Gap in Northwest Himalaya, India. Him Geol 36:153–160

Peltzer G, Saucier F (1996) Present day kinematics of Asia derived from geological fault rates. J Geophys Res 101:27943–27956

Philip G, Virdi NS (2006) Co-existing compressional and extensional regimes along 1764 the Himalayan Front vis-à-vis active faults near Singhauli, Haryana, India. Curr Sci 90:1267–1271

Philip G, Suresh N, Bhakuni SS, Gupta V (2011) Paleoseismic investigation along Nalagarh Thrust: evidence of Late Pleistocene earthquake in Pinjaur Dun, Northwestern subHimalaya. J Asian Earth Sci 40:1056–1067

Philip G, Bhakuni SS, Suresh N (2012) Late Pleistocene and Holocene large magnitude earthquakes along Himalayan Frontal Thrust in the Central Seismic Gap in NW Himalaya, Kala Amb, India. Tectonophysics 580:162–177

Philip G, Suresh N, Jayangondaperumal R (2017) Late Pleistocene-Holocene strain release by normal faulting along the Main Boundary Thrust at Logar in the northwestern Kumaun Sub Himalaya, India. Quatern Int. https://doi.org/10.1016/j.quaint.2017.05.022

Piddington H (1804) Bengal occurrences for October 1803. Asiat Ann Reg 6(35):57–65

Ponraj M, Miura S, Reddy CD, Amirtharaj S, Mahajan SH (2011) Slip distribution beneath the Central and Western Himalaya inferred from GPS observations. Geophys J Int 185:724–736. https://doi.org/10.1111/j.1365-246X.2011.04958.x

Power PM, Lillie RJ, Yeats RS (1998) Structure and shortening of the Kangra and Dehradun re-entrants, Sub Himalaya, India. Geol Soc Am Bull 110:1010–1027

Price RA (1981) The Cordilleran foreland thrust and fold belt in the southern Canadian Rocky Mountains. In: McClay KR, Price NJ (eds) Thrust and Nappe Tectonics. Geol Soc London, Special Publications 9:427–448

Priyanka RS, Jayangondaperumal R, Pandey A, Mishra RL, Singh I, Bhushan R, Srivastava P, Ramachandran S, Shah C, Kedia S, Sharma AK, Bhat GR (2017) Primary surface rupture of the 1950 Tibet-Assam great earthquake along the eastern Himalayan front, India. Nat Sci Reports 7:5433. https://doi.org/10.1038/s41598-017-05644-y

Raiverman V, Srivastava AK, Prasad DN (1994) Structural style in northwestern Himalayan foothills. Him Geol 15:263–280

Rajendran CP, Rajendran K (2005) The status of central seismic gap: a perspective based on the spatial and temporal aspects of the large Himalayan earthquakes. Tectonophysics 395(1–2): 19–39

Rajendran CP, Rajendran K, Sanwal J, Sandiford M (2013) Archeological and historical database on the medieval earthquakes of the central Himalaya: ambiguities and inferences. Seism Res Lett 84(6):1–11. https://doi.org/10.1785/0229130077

Rajendran CP, John B, Rajendran K (2015) Medieval pulse of great earthquakes in the central Himalaya: viewing past activities on the frontal thrust. J Geophys Res Solid Earth 120. https://doi.org/10.1002/2014jb011015

Raper FV (1810) Narratives of a survey for the purpose of discovering the resources of the Ganges. Asiatic Res 11:446–563

Rawat G, Arora BR, Gupta PK (2014) Electrical resistivity cross-section across the Garhwal Himalaya: Proxy to fluid-seismicity linkage. Tectonophysics 637:68–79

Ray Y, Srivastava P (2010) Widespread aggradation in the mountainous catchment of the Alaknanda-Ganga River System: timescales and implications to Hinterland–foreland relationships. Quat Sci Rev 29(17):2238–2260

Robinson DM, Pearson ON (2013) Was Himalayan normal faulting triggered by initiation of the Ramgarh–Munsiari thrust and development of the Lesser Himalayan duplex? Int J Earth Sci 102(7):1773–1790

Robinson DM, DeCelles PG, Copeland P (2006) Tectonic evolution of the Himalayan thrust belt in western Nepal. Geol Soc Am Bull 118:865–885. https://doi.org/10.1130/b25911.1

Sangode SJ, Kumar R, Ghosh SK (1996) Magnetic polarity stratigraphy of Siwalik sequence of Haripur area (HP), NW Himalaya. J Geol Soc India 47:683–704

Sapkota, Rimal (1997) Department of Mining and Geology Annual Report, Nepal.

Sapkota SN, Bollinger L, Klinger L, Tapponnier P, Gaudemer Y, Tewari D (2013) Primary surface ruptures of the great Himalayan earthquakes in 1934 and 1255. Nat Geosci 6:71–76

Sati SP, Sundriyal YP, Rawat GS (2007) Geomorphic indicators of neotectonic activity around Srinagar (Alaknanda basin), Uttarakhand. Curr Sci 92(6):824–829

Satyabala SP, Bilham R (2006) Surface deformation and subsurface slip of the 28 March 1999 Mw = 6.4 west Himalayan Chamoli earthquake from InSAR analysis. Geophys Res Lett 33:L23305. https://doi.org/10.1029/2006GL027422

Schelling D (1992) The tectonostratigraphy and structure of the eastern Nepal Himalaya. Tectonics 11:925–943

Schelling D, Arita K (1991) Thrust tectonics, crustal shortening and the structure of the Far-Eastern Nepal Himalaya. Tectonics 10:851–862

Schiffman C, Bali B, Szeliga W, Bilham B (2013) Seismic slip deficit in the Kashmir Himalaya from GPS observations. Geophys Res Lett 40:5642–5645

Searle MP, Windley BF, Coward MP, Cooper DJW, Rex AJ, Rex D, Li Tingdong, Xuchang X, Jan MQ, Thakur VC, Kumar S (1987) The closing of Tethys and the tectonics of the Himalaya. Geol Soc Am Bull 98(6):678–701

Seeber L, Armbruster JG (1981) Great detachment earthquakes along the Himalayan arc and long term forecasting. In: Simpson DW, Richards PG (eds) Earthquake prediction: an international review. American Geophysical Union, Maurice-Ewing Series, vol 4, pp 259–277

Seeber L, Gornitz V (1983) River profiles along the Himalayan arc as indicators of active tectonics. Tectonophysics 92:335-467. https://doi.org/10.1016/0040-1951(83)90201-9

Sharma VD (2001) Delhi and Its Neighborhood. Archaeological Survey of India, New Delhi, p 161

Sinha AK (1981) Geology and tectonics of the Himalayan region of Ladakh, Himachal, Garhwal-Kumaun and Arunachal Pradesh: a review. In Gupta HK, Delany FM (eds) Zagros, Hidu-Kush, Himalaya Geodynmaic evolution, Geodynamic Series, vol 3. American Geophysical Union, Washington, pp 122–148

Sinha AK (1989) Geology of the part of Higher Central Himalaya. Wiley, p 219

Srivastava P, Mitra G (1994) Thrust geometries and deep structure of the outer and lesser Himalaya, Kumaon and Garhwal (India): Implications for evolution of the Himalayan fold-and-thrust belt. Tectonics 13(1):89–109. https://doi.org/10.1029/93TC01130

Stein RS, King GCP, Rundle JB (1988) The Growth of Geological Structures by Repeated Earthquakes 2. Field Examples of Continental Dip-Slip Faults. J Geophys Res Sol Ea 93 (B11):13319–13331

Stein RS, King GCP, Lin J (1994) Stress triggering of the 1994 M = 6.7 Northridge, California, earthquake by its predecessors. Science 265:1432–1435

Thakur VC (1992) Geology of western Himalaya. Pergamon Press, Oxford, p 366p

Thakur VC (2013) Active tectonics of Himalayan Frontal Fault system. Int J Earth Sci 102(7): 1791–1810

Thakur VC, Kumar S (1994) Seismotectonics of the 20th October 1991 Uttarkashi earthquake in Garhwal Himalaya, North India. Terra Nova 6:90–94

Thakur VC, Pandey AK (2004) Active deformation of Himalayan Frontal Thrust and Piedmont Zone south of Dehradun in respect of seismotectonics of Garhwal Himalaya. Him Geol 25:23–31

Thakur VC, Pandey AK, Suresh N (2007) Late Quaternary-Holocene frontal fault zone of the Garhwal Sub Himalaya, NW India. J Asian Earth Sci 29(2/3):305–319

Thakur VC, Jayangondaperumal R, Malik MA (2010) Redefining Wadia-Medlicott's Main Boundary Fault from Jhelum to Yamuna: an active fault strand of the Main Boundary Thrust in Northwest Himalaya. Tectonophysics 489:29–42

Thakur VC, Joshi M, Sahoo D, Suresh N, Jayangondapermal R, Singh A (2014) Partitioning of convergence in Northwest Sub-Himalaya: estimation of late Quaternary uplift and convergence rates across the Kangra reentrant, North India. Int J Earth Sci 103(4):1037–1056

Thatcher W (1983) Nonlinear strain buildup and the earthquake deformation cycle on the San Andreas fault. J Geophys Res 88:5893–5902

Upreti B, Kumahara Y, Nakata T (2008) Evidence of two large seismic gaps in Nepal Himalaya: potential for future mega earthquakes. 33rd International Geological Congress, Oslo. http://www.cprm.gov.br/33IGC/1260752.html

Upreti BN et al (2000) The latest active faulting in Southeast Nepal. Paper presented at Proceedings active fault research for the New Millenium. Hokudan international symposium and school on active faulting, Awaji Island, Hyogo, Japan

Valdiya KS (1980) Geology of Kumaun Lesser Himalaya. Wadia Institute of Himalayan Geology Dehradun India, p 219

Valdiya KS (2003) Reactivation of Himalayan Frontal Fault: Implications. Curr Sci 85(7):1031–1040

Valdiya KS (1992a) Active Himalayan Frontal Fault, Main Boundary Thrust and Ramgarh Thrust in southern Kumaun. Geol Soc India 40:509–528

Valdiya KS (1992b) The Main Boundary thrust zone of the Himalaya, India. In: Bucknam RC, Hancock PL (eds) Major active faults of the world: results of IGCP Project 206: annales tectonicae supplement, vol 6, pp 54–84

Vassallo R, Vignon V, Mugnier J-L, Jayangondaperuma, R, Srivastava P, Malik MM, Mouchené M, Jouanne F (2012) A future big one in the Himalayan North-West syntax? European Geosciences Union General Assembly, Vienna, Austria 22–27 Apr 2012

Vassallo R et al (2015) Distribution of the Late Quaternary deformation in Northwest Himalaya. Earth Planet Sci Lett 411:241–252

Wells D, Coppersmith K (1994) New empirical relationships among magnitude, rupture length, rupture width, rupture area, and surface displacement. Bull Seismol Soc Am 84:974–1002

Wesnousky SG, Kumar S, Mahindra R, Thakur VC (1999) Uplift and convergence along the Himalayan Frontal Thrust of India. Tectonics 18:967–976

Yeats RS, Thakur VC (2008) Active faulting south of the Himalayan Front: establishing a new plate boundary. Tectonophysics 453:63–73

Yeats RS, Nakata T, Farah A, Mirza MA, Pandey MR, Stein RS (1992) The Himalayan Frontal Fault system. In: Bucknam RC, Hancock PL (eds) Major active faults of the world: results of IGCP Project 206: annales tectonicae, supplement 5, 6, pp 85–98

Yin A (2000) Mode of Cenozoic eastwest Extension in Tibet suggests a common origin of rifts in Asia during Indo- Asian collision. J Geophys Res 105:21745–21759

Yin A (2006) Cenozoic tectonic evolution of the Himalayan orogen as constrained by along–strike variation of structural geometry, exhumation history, and foreland sedimentation. Earth Sci Rev 28:211–280

Yu S-B, Kuo L-C, Hsu Y-J, Su HH, Lui C-C, Hou C-S, Lee J-F, Lai T-C, Liu C-C, Liu C-L, Tseng
    T-F, Tsai C-S, Shin T-C (2004) Preseismic deformation and coseismic displacements
    associated with the 1999 Chi-Chi, Taiwan, Earthquake. Bull Seism Soc Am 91(5):995–1012
Yule D, Dawson S, Jerome Lave, Sapkota S, Tiwari D (2006) Possible evidence for surface
    rupture of the Main Frontal Thrust during the great 1505 Himalayan earthquake, far-western
    Nepal, Eos Trans. AGU, 87(52), Fall Meet. Suppl., Abstract S33C-05
Zhao W, Nelson KD, Project INDEPTH Team (1993) Deep seismic reflection evidence for
    continental underthrusting beneath southern Tibet. Nature 302:557–559

Chapter 2
Mapping of Active Faults

2.1 Introduction

The mapping of fault scarp is of great interest to understand the neo-tectonic process present in the region. A fault scarp is defined as tectonic landforms corresponding, or roughly coincident, with a fault plane that has displaced the ground surface (e.g., 1994 Northridge, 1999 Chi-Chi and 2005 Kashmir earthquakes). Investigations on fault upliftment, fold axis, stress field and plate motion provide present stage of crustal deformation and possibility of recurrence of earthquake in the Himalayas (Gansser 1964; Nakata et al. 1990; Yeats and Thakur 1998; Wesnousky et al. 1999; Lave and Avouac 2000; Kumar et al. 2001; Banerjee and Burgmann 2002; Jayangondaperumal et al. 2011; Mugnier et al. 2013; Avouac et al. 2015). Several researchers have used remote sensing data for mapping of active faults based on on-screen visual interpretation and geomorphic variation such as upliftment, stream offsetting, incised meandering, and tilting terraces (e.g., Nakata and Kumahara 2002; Nakata et al. 1991; Harding and Berghoff 2000; Philip 2007; Hilley et al. 2010). Stereopairs, LIDAR and SAR interferometry are advanced imageries offer bare-earth surface model to identify active faults present in the region. The present work is constituted by integration of bare-earth surface model and geomorphic surface features to map the active faults. The subsequent section described basics of remote sensing techniques which will be useful to understand the techniques and processes involved in the active fault mapping.

2.1.1 Basics of Remote Sensing Techniques

Remote sensing is a widely used optical recognition technique for characterising the earth surface process and landforms without touch. When electromagnetic radiation interacts with matter, light energy is partially absorbed, reflected, and transmitted.

© Springer Nature Singapore Pte Ltd. 2018
R. Jayangondaperumal et al., *Active Tectonics of Kumaun and Garhwal Himalaya*,
Springer Natural Hazards, https://doi.org/10.1007/978-981-10-8243-6_2

Critical analyses of these energies across optical wavelength regions provide an excellent scope for understanding the materials composition. In common, remote sensing sensors are operated in two different platforms, namely airborne (i.e., aircraft) and spaceborne (i.e., satellites). Based on the utilization of light source, the sensors can be classified either active or passive. Active sensors have their own source of energy, whereas passive sensors utilized sun radiation as a source for earth observation (Lillesand et al. 2011). During the data acquisition process, remote sensing sensors record the radiance value of the object and stored it by digital counts. These digital counts can be utilized for spatial analysis as well as spectral analysis. The magnitude of spatial and spectral response of the object at any specific wavelength has been related to the abundance of constituent material within the field of view (FOV).

2.2 Data and System

2.2.1 Cartosat—1A Stereopair Data

Cartosat—1A (IRS—P5) is the first Indian stereoscopic remote sensing sensor launched by Indian Space Research Organization (ISRO) on May 5, 2005. The imagery has stereoscopic earth observation sensor which provides pair image products for digital terrain analysis. Each image pixel has spatial resolution of 2.5 m, covered wide spectra of visible region (500–850 nm). One of the image pairs (Band A) has off-nadir angle of $-5°$ and another one (Band F) has off-nadir angle of $+26°$. In common, Band A is used for left eye perspective and Band F used for right eye perspective. The fore (Band F) and aft (Band A) cameras scanned ground surface through charge coupled detector (CCD) array and stored the data in GeoTIFF format. GeoTIFF is a widely used raster file format which is platform independent and has capable to store geographic information together with image data.

The Orthokit data product includes stereopair image along with metadata and rational polynomial coefficients (RPC). The RPC values provide true ground control points that can be used for image triangulation and Orthorectification. Cartosat—1A orthokit product is available at the National remote sensing centre (NRSC), Hyderabad. Specifications of Cartosat—1A stereopair data used in the present study are presented in Table 2.1.

Table 2.1 Specifications of Cartosat—1A stereopair data used in the present study

S. No	Path/ row	Date of acquisition	Latitude (°N)	Longitude (°E)	Pixel resolution (m)	Ellipsoid and Datum	Swath width (km) (F: Fore/A: Aft)
1	525/257	26-Oct-2006	30.575831	77.773793	2.5	WGS_84	F: 29/A: 26
2	525/258	26-Oct-2006	30.351751	77.717436	2.5	WGS_84	F: 29/A: 26
3	526/258	5-Feb-2006	30.351785	77.913000	2.5	WGS_84	F: 29/A: 26
4	526/259	5-Feb-2006	30.127808	77.856292	2.5	WGS_84	F: 29/A: 26
5	526/261	26-Mar-2009	29.681175	77.750403	2.5	WGS_84	F: 29/A: 26
6	527/258	15-Mar-2009	30.351425	78.107374	2.5	WGS_84	F: 29/A: 26
7	527/259	15-Mar-2009	30.127445	78.050251	2.5	WGS_84	F: 29/A: 26
8	527/260	15-Mar-2009	29.903446	77.993319	2.5	WGS_84	F: 29/A: 26
9	527/261	30-Mar-2011	29.679133	77.958821	2.5	WGS_84	F: 29/A: 26
10	528/259	2-Nov-2010	30.127358	78.300496	2.5	WGS_84	F: 29/A: 26
11	528/260	2-Nov-2010	29.903357	78.243895	2.5	WGS_84	F: 29/A: 26
12	529/260	20-Oct-2009	29.905808	78.368625	2.5	WGS_84	F: 29/A: 26
13	529/261	20-Oct-2009	29.681768	78.312439	2.5	WGS_84	F: 29/A: 26
14	530/260	9-Oct-2009	29.905758	78.555627	2.5	WGS_84	F: 29/A: 26
15	530/261	9-Mar-2012	29.679320	78.516273	2.5	WGS_84	F: 29/A: 26
16	531/261	9-Nov-2008	29.681183	78.705676	2.5	WGS_84	F: 29/A: 26
17	531/262	24-Oct-2011	29.457637	78.658160	2.5	WGS_84	F: 29/A: 26
18	532/261	21-Jan-2010	29.681710	78.940134	2.5	WGS_84	F: 29/A: 26
19	532/262	21-Apr-2014	29.455421	78.812703	2.5	WGS_84	F: 29/A: 26
20	533/261	30-Dec-2009	29.679388	79.058821	2.5	WGS_84	F: 29/A: 26
21	533/262	30-Dec-2009	29.455352	79.003337	2.5	WGS_84	F: 29/A: 26
22	534/262	17-Dec-2008	29.457493	79.233788	2.5	WGS_84	F: 29/A: 26
23	534/263	22-Apr-2009	29.231284	79.166682	2.5	WGS_84	F: 29/A: 26

(continued)

Table 2.1 (continued)

S. No	Path/row	Date of acquisition	Latitude (°N)	Longitude (°E)	Pixel resolution (m)	Ellipsoid and Datum	Swath width (km) (F: Fore/A: Aft)
24	535/262	19-Dec-2009	29.457997	79.402785	2.5	WGS_84	F: 29/A: 26
25	535/263	19-Dec-2009	29.233910	79.347007	2.5	WGS_84	F: 29/A: 26
26	536/263	8-Dec-2009	29.234072	79.627023	2.5	WGS_84	F: 29/A: 26
27	536/264	8-Dec-2009	29.009969	79.571257	2.5	WGS_84	F: 29/A: 26
28	537/264	9-Mar-2009	29.007188	79.678055	2.5	WGS_84	F: 29/A: 26
29	538/263	3-Nov-2008	29.233016	79.948769	2.5	WGS_84	F: 29/A: 26
30	538/264	3-Nov-2008	29.008933	79.892936	2.5	WGS_84	F: 29/A: 26
31	539/263	26-Feb-2009	29.231299	80.127207	2.5	WGS_84	F: 29/A: 26
32	539/264	26-Feb-2009	29.007209	80.071454	2.5	WGS_84	F: 29/A: 26

2.2.2   System Configuration and Software Setup

The size of each Cartosat—1A stereopair image is 0.5 GB and therefore processing and analysis of the data need advanced hardware and software setup with large internal/external memory storage. The hardware and software accessories used in this purpose are presented in Note 2.1. The workstation (computer system) with dual monitors was fixed at the Structure and tectonics laboratory of Wadia Institute of Himalayan Geology, Dehradun and SOCETGXP 4.0 software was configured in that system with the help of SOCET installation manual.

2.3   Mapping Techniques

The primary goal of this chapter is to provide complete processing steps to make this book self-contained. The methodology adopted in this research has been divided into three distinct phases: (1) structure and tectonics of the Kumaun and Garhwal Himalaya, (2) possible locations to identify active faults, and (3) active fault mapping using remote sensing techniques. The integrated methodological framework is presented in Fig. 2.1. Discussions made on first and second phases have been already done in the first chapter (see Sect. 1.11). Henceforth, detailed methodology of phase three is described in the subsequent sections.

2.3.1   Image Calibration

The raw orthokit products have pixel distortion and shrinkage because imaging sensors failed to capture undulated surface in the elevated area. Therefore, image calibration is a prerequisite to characterize actual earth surface present in the region (Floyd 2007). This process has been done by the triangulation module of SocetGXP software. Triangulation module provides a relative orientation of the image plane and converts the tilted photographic system into equivalent vertical photographic system. In the primary situation, stereopair images were imported into software workspace and display it in the separate window. The relative coverage between two pairs has been cross-checked by the landmark points (e.g., road corner, buildings, foothill edge). After crucial evaluation, the images were imported into triangulation model for Orthorectification process. The module has inbuilt function to read sensor type, model parameters, accuracies, covariance data and constraints from the metadata of the image pairs. These parameter values were fixed as reference constraints and further rectification has been done by automatic tie points and ground control points (GCP).

Work flow of Triangulation module includes image setup, data setup, automatic point measurement (APM), interactive point measurement (IPM), and solve tab. The image setup tab consists of selected image files along with metadata. The

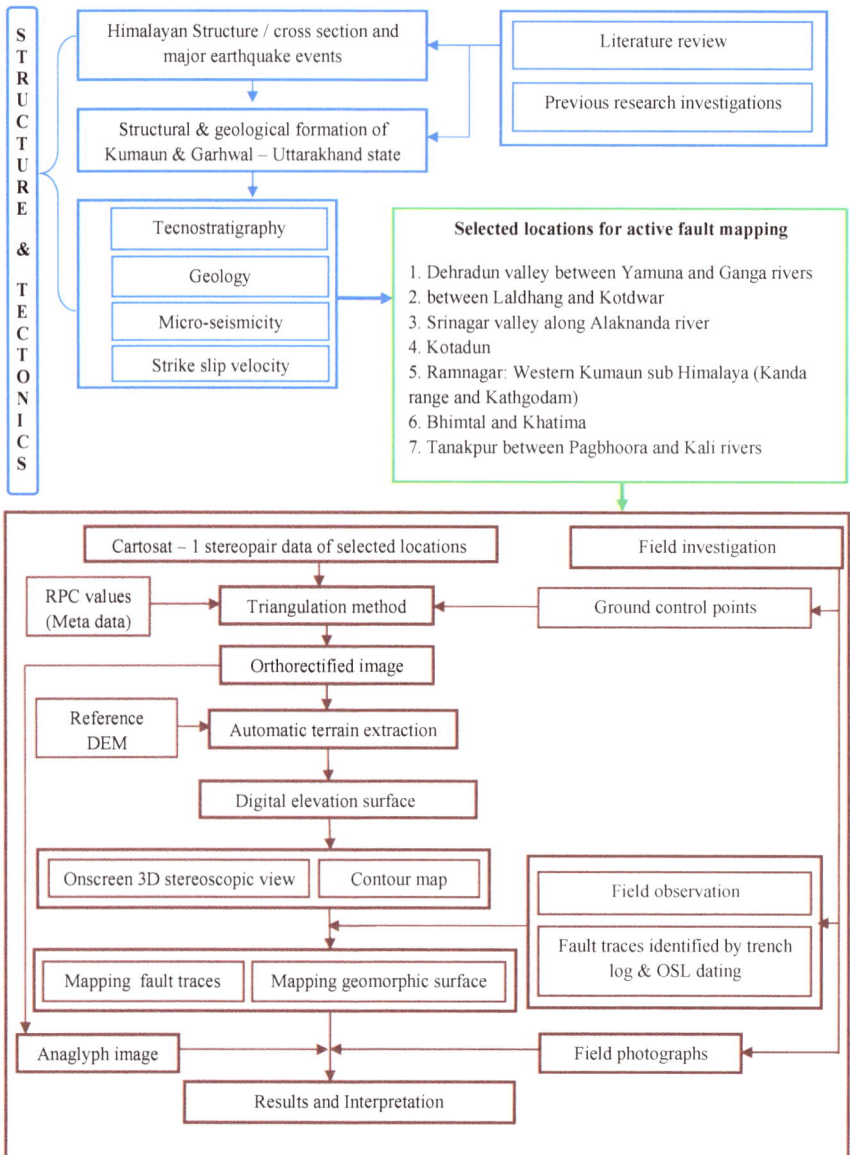

Fig. 2.1 Methodology flowchart showing process involving in the active fault mapping

sensor type and calibration details are populated based on the PRC file and displayed in the image data section. The data setup tab is used to add feed datasets like ground control points and existing digital elevation files (in this case, SRTM 30 m resolution is used (source: http://earthexplorer.usgs.gov/). These datasets assist automatic point measurement (APM) to derive actual elevation which is helpful to increase the accuracy of the triangulation results. The APM tab includes different algorithm strategies, automatic tie point pattern, and result analysis tab. The present case, we used automatic point measurement algorithm with even pattern of 125 tie points. The result in triangulation overview shows automatic selection of tie points in the image pairs.

Note 2.1

Hardware and software specifications

S. no	Hardware	Configuration
1	Processor	Intel Xeon processor E5-1607 V3 @ 3.10 GHz/core i7
2	Chipset	Intel C610
3	RAM	64GBDDR4/8 RDIMMECC slot/2133 MHz
4	Storage	2 X 4 TB SATA (7200 rpm)
5	Graphics card	NVIDIA QuadroK5200
6	Monitor	Dual monitor a) BenQ (1920 X 1080 resolution with 120 Hz illumination) 3D monitor b) DELL (1920 X 1080 resolution) monitor
7	3D Glass	NVIDIA 3D Vision wireless kit
8	Operating system	Genuine Windows 8.1 Pro (64 bit)

S. no	Software	Features
1	SOCETGXP 4.0	The SOCETGXP 4.0 includes • Workspace manager • Multiport image analysis toolbox • Triangulation module • Mosaic/transform module • Automatic terrain extraction module • Terrain post-processing toolbox • 3D contour and terrain visualization • Automatic feature extraction module • 3D/2D shape file creation module • 3D building toolbox • Export vector/raster/DEM data

The solve tab executes the strategy and provides the accuracy status by RMS residual error. The acceptability of the RMS residual error is less than 0.30 pixels. Once the process is complete, interactive point measurement (IPM) multiport is used to graphically see the bad image points and correct or eliminate them manually. Iterate this process up to satisfactory RMS residual error. The entire process is stored in triangulation file format (*.arf) so that it can be used for further modifications. The example view of stereopairs before and after triangulation process is presented in Note 2.2. Once the triangulation is complete, images with calibrated information stored in the socetGXP support file (*.sup) format.

2.3.2 Digital Elevation Model (DEM)

Automatic Terrain Generation (ATG) module is a SocetGXP geospatial intelligence tool that has been used in the present study to extract terrain information from the triangulated stereoscopic imagery. It uses three types of intelligence algorithms, namely Next Generation Automatic Terrain Extraction (NGATE), Automatic Terrain Extraction (ATE), and Automatic Spatial Modeler (ASM). Among this, NGATE outperforms image correlation and edge matching on each image pixel and is best used for difficult areas such as large-scale imagery in urban and hilly terrain areas. It provides elevation surface either a bare-earth surface (digital terrain model (DTM)) or digital surface model (DSM).

The bare-earth DTM model removes tree and building structures based on the minimum height and the maximum width, smoothing, precision, TIN breakline, and TIN masspoints. The constraint of minimum height and maximum width was fixed with reference to the existed digital elevation surface (SRTM 30 m DEM). Therefore, buildings or trees with greater than constraint values are automatically eliminated. The ATG window provides option to the users to create digital elevation based on their requirements. The present case, we preferred Grid surface model, DTM resolution of 5 m (along and across spacing), and resolution unit in meters. The parameter module includes three set of parameters namely image, strategy, and seed DTM. The image tab contains the name and path of the triangulated images used to run ATG. In the strategy tab, we preferred NGATE strategy because it can reduce post-processing time and has provided accurate DTM than others. Constraint parameters like minimum height and maximum width are fixed as 10 and 60 m, respectively. Moreover, high smoothing and precision and slow processing speed are fixed for DTM creation. In the seed DTM tab, we supplied existed SRTM 30-m digital terrain surface. The result of ATG process shows finest bare-earth surface. However, post-processing is employed to remove unwanted peaks and flatten the smooth surface.

2.3.3 Post—Terrain Process

The DTM post process helps to remove unwanted peaks or structures present in the bare-earth surface model. The SocetGXP class algorithm tool is used to assign an algorithm for different type of corrections. The tool includes intelligent polygon-based algorithms, namely fill, bare-earth, interpolate, bias, clip, thinning, and visibility. Through on-screen visual investigation, sudden peaks and down were identified in the DTM data and removed by the fill process. Conversely, the bare-earth areas with minimal visible ground points such as dense forest and cluster of buildings were removed by the bare-earth algorithm, whereas clustering of tiny surface objects was removed by the interpolate algorithm. The bias algorithm is utilized to adjust the elevation of each grid cell based on the existing ground control points. The processed DTM data are stored in Socet elevation data (*.dth) format.

Note 2.2

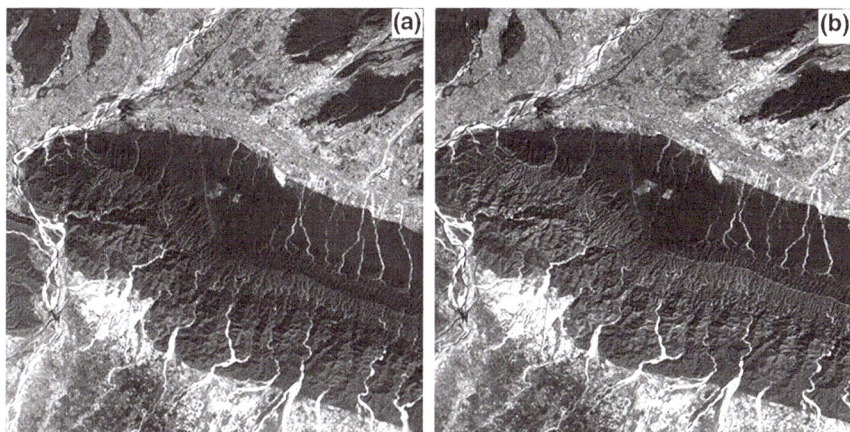

Raw Cartosat—1 stereopair image (**a** Band F, **b** Band A) (*Location* Mohand, NW Himalaya)

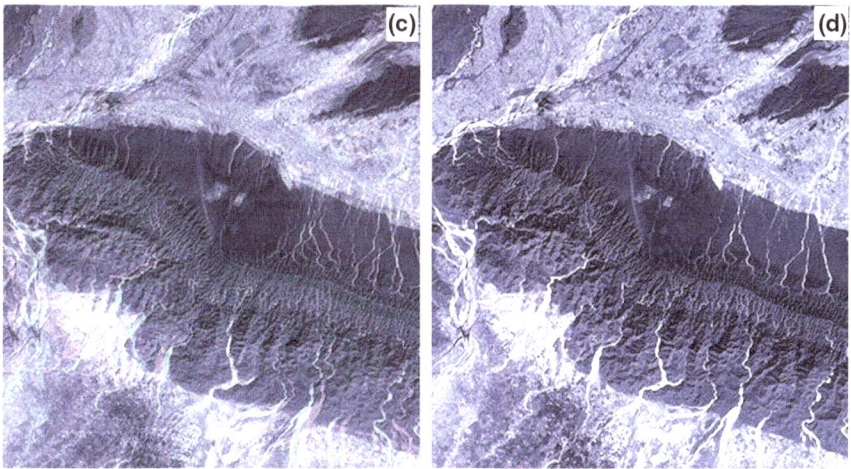

Band A overlaid on Band F with transparency of 40% (**c** before calibration, **d** after calibration)

2.3.3.1   Quality Statistics

The quality statistics provides quality assurance statistics on the differences between DTM data and reference dataset. The files can be any combination/spatial resolution of terrain data, e.g., two terrain files, two ground control point files, or a ground point file to a terrain file. In the present case, DTM data are compared with existed SRTM 30 m data as well as ground control points (GCP). The report editor in SocetGXP is used to edit and annotate the quality statistics report. The result contains the type of quality statistics checked, the name of the files compared, and results of the statistical comparisons [mean, standard deviation, and root mean square error (RMSE)]. The DTM data with RMSE error less than 0.10 are considered for further operations. Note 2.3 exhibits quality of Cartosat—1A DTM 5 m resolution as compared to the SRTM 30 m resolution.

Note 2.3
Terrain accuracy of SRTM 30 m and present model
   (Cartosat—1A DTM 5 m)
Elevation range: Min: 475; Max: 1878
Location: Dun surface between Dungakhet and Tons nadi

SRTM 30 m resolution

Cartosat—1 DTM with 5 m resolution

2.3.3.2 Terrain Operation

The terrain operation module in SocetGXP helps to perform merging multiple terrain files into a single file, splitting a single terrain file into multiple terrain file and changing the terrain file format. The tool is used here for transform the DTM original file format (*.dth) into the common terrain file format (GeoTIFF). DTM data with GoeTIFF file format can also be used in other geospatial software like ArcGIS, Erdas imagine, ENVI, global mapper.

2.3.4 Anaglyph Image Production

Anaglyph images are 3D images that can be viewed in a 2D monitor with the help of 3D glasses (a pair of red and cyan glasses). The two raster images overlap with accepted parallax angle of eyesight provide 3D perspective of the objects present in the image. In the present case, anaglyph images help to locate fault traces, geomorphic surface break and river terraces by visual interpretation method. The Cartosat—1 stereopair sensor captured the scene at specific parallax angle so that anaglyph images can be directly derived from the Orthorectified stereopair images without complication. There are several software tools and tutorials which are available for making anaglyph images. Among them, few significant tools were discussed. The triangulation process provides right and left epipolar images that have been utilized for making anaglyph images. The SocetGXP has inbuilt function of "color composite" to make the anaglyph image instantly (Note 2.4 top right). The Integrated Land and Water Information System (ILWIS—version 3.12) is GIS software which contains visualization toolkit for anaglyph creation. In this toolbox, select the stereo Orthorectified images in the catalog, right click on visualization tool, select anaglyph, and use the Red–Blue option. The system monitor displays anaglyph image of the stereopair data. Similar to that, ENVI version 4.7 is a geospatial tool can generate anaglyph images using DEM Extraction wizard. The wizard used tie points to calculate epipolar geometry of the stereopair data. The epipolar reduction factor is an inbuilt comment to control the down sampling of right and left epipolar images. The "Examine Epipolar results" section helps to create 3D epipolar (anaglyph) image by selection of RGB color channels. The left epipolar image is selected for red channel, whereas right epipolar image selected for blue and green, respectively. The combined RGB mode creates anaglyph image that can be viewed by 3D anaglyph glasses.

Anaglyph images can also be created from the non-GIS software like MATLAB, and Adobe Photoshop. In the MATLAB environment, the Orthorectified stereo images can be imported by "geotiffread" inbuilt command. It stores geocoordinates and map projections as a structured parameters that will be directly utilized in GIS software without georectification. The stereopair data individually read by two assigned parameters (e.g., I: left image and J: right image) and rectify it by using the inbuilt command "[X, Y] = rectifyStereoImages (I, J, stereoParams)." StereoParam is consisting of intrinsic and extrinsic parameters of the stereo cameras. Next to that, "A = stereoAnaglyph (X, Y)" inbuilt command is used to create the anaglyph images. The "geotiffwrite" inbuilt command helps to create "GeoTIFF" image of output anaglyph along with the geocoordinates and map projections. Conversely, Photoshop is the powerful image analysis software that has capable of generating anaglyph images from the stereopair data. As a primary step, stereo images need to be imported to the Photoshop layer stack. Select the Band A (left epipolar) image from the layer panel, choose the blending option. Further, go to the advanced blending menu, select R channel (i.e., red channel) and deselect G and B channels; click OK button. Similarly, select the B and F (right epipolar) image from the layer, go to advanced blending menu, select G and B and deselect R. As a result, the Photoshop image window displays an anaglyph image of the stereopair data. However, Photoshop software reads only pixels values and it does not consider geocoordinates and map projections of the stereopair data.

Anaglyph shows three-dimensional surface features in a two-dimensional image frame. This will be commonly used to identify the active faults and geomorphic features present in the region. However, it has certain limitations; the one mainly is lack of multi-dimensional rotation and the other one is not viable to expand vertical exaggeration with respect to the convenient prospective view. To overcome these limitations, we proposed a simple technique for 3D visualization of anaglyph images. The first step is to create anaglyph images using one of the techniques as mentioned above. Further, overlay anaglyph image on the existed DTM elevation surface. The difference between normal and 3D anaglyph view is presented in Note 2.4.

2.3.5 Contour Line Extraction

Contour lines are derived from the Digital Terrain Model which exposes terrain features as a line of constant elevation interval. It helps to identify the surface deformation based on the closeness of the contour interval. In the present study, contour lines are drawn by using SocetGXP geospatial module. The elevation base is fixed to the mean sea level (MSL) and interval between contour lines is fixed at 5 or 10 m. The derived contour lines are stored as vector data format (shapefile, *. shp). Note 2.5 shows significance between normal DTM and contour overlaid DTM.

2.3.6 Geomorphic Surface Mapping

The integration of Orthorectified image, DTM, and contour lines helps to identify geomorphic surfaces present in the region. In the present work, geomorphic surfaces are mapped by on-screen visual interpretation based on the pixel tone, topography, texture, and contour density (Table 2.2). The geomorphic features come under the following six categories namely Dun surface, Piedmont plain, river terraces, alluvial plain, floodplain/channel bar, and Siwalik hills. Siwallik hills in the Kumaun and Garhwal Himalaya are appeared as dark tone, rough texture, and dense vegetated cover. The southern part of the Siwalik foothills is dominated by piedmont fans of varying dimensions, which delineates the paleochannel alluvial deposits. Topography of the piedmont surface shows smooth crest ridge, dense vegetation and terrace upliftment. River terraces are appeared at adjacent to the present river bed. Depending on the existence of paleochannel environment, different levels of terraces are exposed in the river bed (Pathak et al. 2015). Dun surfaces are originated by the exhumation of frontal Siwalik belt along the Himalayan Frontal Thrust (HFT). These Duns are appeared as gentle slope which are mostly formed by major drop in stream gradient at the HFT. Alluvial plain surface and floodplain/channel bars are located at the downstream areas which are mostly of smooth, flat terrain, and discontinuous texture.

Table 2.2 Geomorphic features and its spatial characteristics

S. No	Feature	Tone	Topography	Texture	Contour
1	Siwalik hills	Dark tone due to dense vegetation	Bedding traces, strike ridge and valley and steep slope	Rough texture	High closeness of contour traces
2	Dun surface	Light tone due to settlement and dark tone due to dense vegetation	Smooth terrace and gentle sloping nature	Fan shape	Gradual variation of contour traces
3	Piedmont plain	Moderately dark due to dense vegetation	Smooth crest ridge surface with vegetation	Smooth convex texture	Smooth closeness of contour traces
4	River terraces	Light tone	Adjacent to the present river bed with flat surface	Smooth texture with cut and strath terraces	Diverge at terrace surface and closeness at terrace edge
5	Alluvial plain	Moderately light tone due to agriculture land and dark tone due to dense vegetation	Wide span of flat surface with vegetation	Smooth texture	Diverge contour traces
6	Floodplain/ channel bar	Dry sand gives light tone and wet sand gives dark tone	Nearly flat surface with vegetation/sand bed	Smooth discontinued texture	Vast diverge contour traces

Note 2.4
Anaglyph image

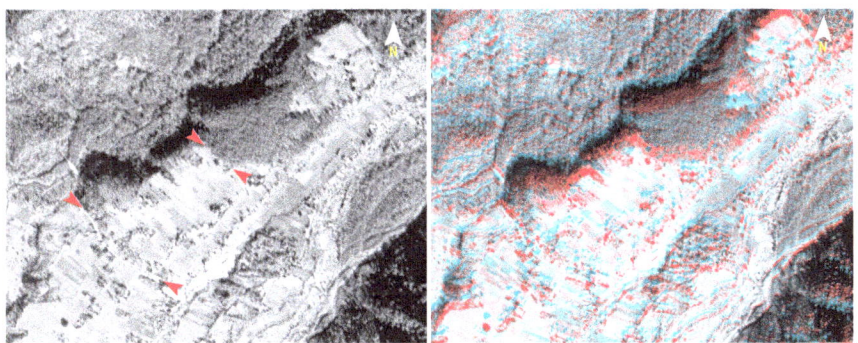

Cartosat—1A grayscale image (left) and anaglyph image (right). Red arrow indicates fault scarp (*Location* Kotabagh)

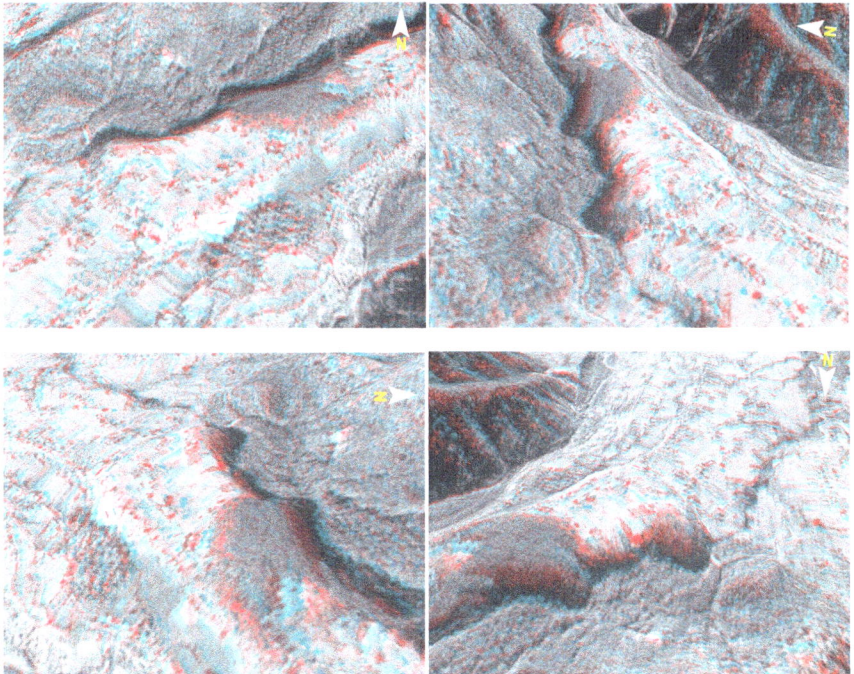

3D anaglyph shows fault scarp view at different directions (In clockwise, north, east, south, and west)

Note 2.5
Significance of Contour lines

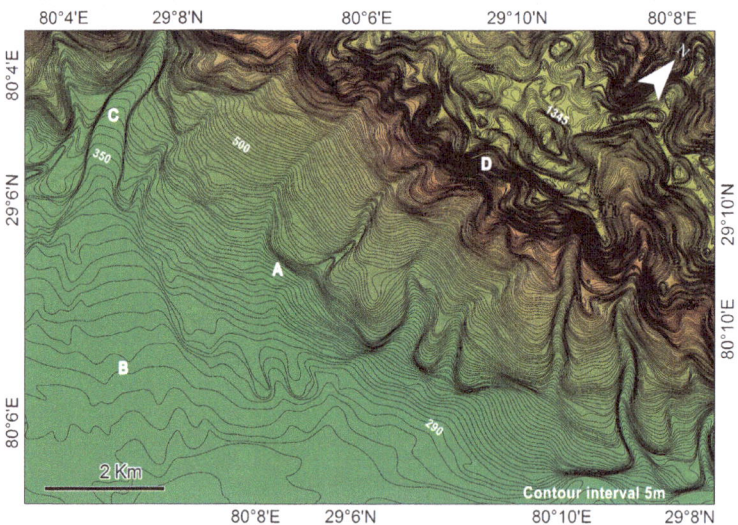

Cartosat—1 DTM with 5 m resolution without contour lines (provide no significant information)

Contour lines with 5 m interval overlay on Cartosat—1 DTM (Location: Tanakpur). A: sudden closeness of contour lines reveals active faults at the piedmont fan surface, B: gradual increase in contour lines shows recent alluvial fan surface, C: contour lines with inverted "V" shape shows present river bed, D: closed or concentric contour lines show hilly terrain or peaks

2.3.7 Active Fault Mapping

Active faults present in the region are identified by geomorphic fluctuation and closeness of contours. A fault scarp is defined as tectonic landforms corresponding, or roughly coincident, with a fault plane that has displaced the ground surface. In order to map the fault scarp efficient way, few general principles are considered, i.e., (1) sudden upliftment is identified at the downhill region that can be south or north facing, (2) deep cut in river terraces that is perpendicular to the present river bed, (3) active fault traces may inferred at the southern part of mountain foothill due to the long period of fluvial modification. Moreover, differential uplift of the river terraces and past evidence of channel shift are also signatures of neo-tectonic activities (Nakata et al. 1991; Nakata and Kumahara 2002; Jayangondaperumal et al. 2011, 2017; Malik and Nakata 2003; Pandey et al. 2014). The identified fault scarps are mapped systematically by using ArcGIS 10 software. Fault scarp properties are prepared in a table format that contain thrust name, fault scarp type, geocoordinate information, strike direction, strike—slip length, affected geomorphic surface and location. In addition to that, fault traces are shown in the anaglyph 3D stereoscopic view which is appended in the Annexure section. All the mapping features such as geomorphic surface, faults, and its attribute information (vector shapefile) are also projected in the Web interface system.

2.4 Active Fault Characterizations Using Paleoseismogical Investigations

The Himalayan frontal fault system is the longest and fastest moving continental convergence system, with the potential for producing devastating, large-magnitude earthquakes. The Indo-Gangetic fertile plains lie along the HFT at the base of the Sub-Himalaya, where large populations live and farm. Establishing the timing of earthquakes along the HFT is critical in assessing regional seismic hazards for these areas proximal to the fault that have both large populations and potentially inadequate infrastructures. However, determining the seismic hazards in this region has been difficult due to uncertainties in basic fault parameters, including the locations and rupture lengths of historical earthquakes, earthquake recurrence intervals, segmentation, and co-seismic slip. Each of these parameters can be established through mapping of active faults and paleoseismic investigation and can then be used to generate predictive, probabilistic models for both the magnitude and likelihood of future earthquakes.

Fault scarp is the fault-generated landforms and it is sensitive indicators of the style and timing of tectonic/earthquake activity (Note 2.6). Fault scarp is defined as the tectonic landform coincident with a fault plane that has dislocated the ground surface (Note 2.6). Trenching across the fault scarp provides valuable information for seismic hazard assessment of the given region. Various types of fault scarps exist along the active fault system and it tells us the near-surface deformation and duration of faulting. In the Himalaya along the foothills, there are several fault scarps that look like a baby mountain in front of the Sub-Himalayan frontal fault system that grew like a hill and then mountain due to repeated earthquake activity and thus mountain-building process.

2.4.1 Paleoseismology

Trench is dug perpendicular to the mapped active fault and logs the successive ruptures in cross section (Note 2.7). Mapping the various structures (faults, folds, sediments layers) and sediment units in the dugged trench helps to identify both the stratigraphy of the sedimentary layers and the primary surface ruptures of the earthquake. Dating of organic materials in the different sedimentary units provides absolute ages by AMS and classic ^{14}C dating methods. The absolute ages of stratigraphic records will be transferred into the calendar time frame and thus timing of seismic event visible in the trench exposure ultimately provides the paleoseismic history. Sediment samples are also collected for dating of stratigraphic units using Optically Stimulated Luminescence (OSL) technique to constrain the depositional age of the unit or to provide a firm lower bound of the earthquake event/s, and also for comparison of ages obtained by other methods (e.g. radiocarbon or fall-out isotopes dating).

Note 2.6

Field photograph shows fault scarp at the base of the Sub-Himalaya (view from south). Field site is located along the mountain front between Belparao and Kaladungi foothills village, Kumaun Himalaya

Note 2.7

Pre-Earthquake Surface

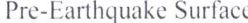

Earthquake produced Fault Scarp

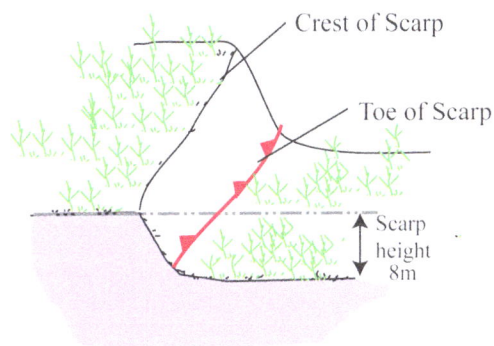

Trench excavation across the fault scarp

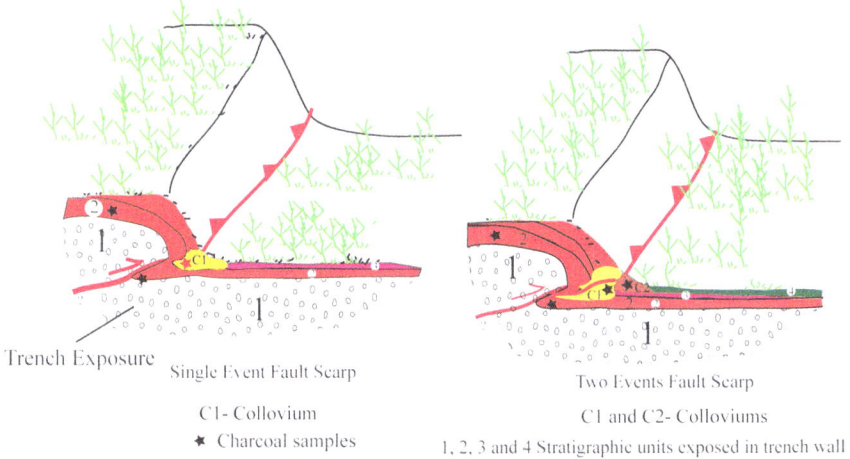

Single Event Fault Scarp

C1- Collovium

★ Charcoal samples

Two Events Fault Scarp

C1 and C2- Colloviums

1, 2, 3 and 4 Stratigraphic units exposed in trench wall

A simplified cartoon illustrates trench investigation across the fault scarp

References

Avouac Jean-Philippe, Meng Lingsen, Wei Shengji, Wang Teng, Ampuero Jean-Paul (2015) Lower edge of locked Main Himalayan Thrust unzipped by the 2015 Gorkha earthquake. Nat Geosci 8(9):708–711

Banerjee P, Burgmann R (2002) Convergence across the Northwest Himalaya from GPS measurements. Geophys Res Lett 29 (13):30-1–4. https://doi.org/10.1029/2002gl015184

Floyd FS (2007) Remote sensing: principles and interpretation, 3rd edn. Waveland Pr Inc. 464p

Gansser A (1964) The geology of the Himalaya. Wiley Interscience, New York, p 189

Harding DJ, Berghoff GS (2000) Fault scarp detection beneath dense vegetation cover: airborne lidar mapping of the seattle fault zone, bainbridge island, Washington state. In Proceedings of the American Society of Photogrammetry and remote sensing annual conference, Washington, D.C., May 2000

Hilley GE, DeLong S, Prentice C, Blisniuk K, Arrowsmith JR (2010) Morphologic dating of fault scarps using airborne laser swath mapping (ALSM) data. Geophys Res Lett 37:L04301. https://doi.org/10.1029/2009GL042044

Jayangondaperumal R, Wesnousky SG, Choudhuri BK (2011) Near-surface expression of early to Late Holocene displacement along the Northeastern Himalayan Frontal Thrust at Marbang Korong Creek, Arunachal Pradesh, India. Bull Seism Soc America 101(6):3060–3064

Jayangondaperumal R, Kumahara Y, Thakur VC, Kumar A, Srivastava P, Shubhanshu D, Joevivek V, Dubey AK (2017) Great earthquake surface ruptures along back thrust of the Janauri anticline, NW Himalaya. J Asian Earth Sci 133:89–101. https://doi.org/10.1016/j.jseaes.2016.05.006

Kumar S, Wesnosusky SG, Rockwell TK, Ragona D, Thakur VC, Seitz GG (2001) Earthquake recurrence and rupture dynamics of Himalayan Frontal Thrust, India. Science 294:2328–2331

Lave J, Avouac JP (2000) Active folding of fluvial terraces across the Siwaliks Hills, Himalayas of Central Nepal. J Geophys Res 105(B3):5735–5770

Lillesand, Kiefer, Chipman (2011) Remote sensing and image interpretation, 6th edn. Wiley India Pvt Ltd. 772p

Malik JN, Nakata T (2003) Active faults and related Late Quaternary deformationalong the northwestern Himalayan Frontal Zone India. Ann of Geophys 46(5):917–936

Mugnier JL, Gajurel A, Huyghe P, Jayangondaperumal R, Jouanne F, Upreti B (2013) Structural interpretation of the great earthquakes of the last millennium in the central Himalaya. Earth Sci Rev 127:30–47

Nakata T, Otsuki K, Khan SH (1990) Active faults, stress field, and plate motion along the Indo-Eurasian plate boundary. Tectonophysics 181:83–95

Nakata T, Khan SH, Lawrence RH (1991) Active faults of Pakistan, Research Center for Regional Geography, Hiroshima University, Hiroshima, Special Publication, 21, ISBN 4-938580-05-5

Nakata T, Kumahara Y (2002) Active faulting across the Himalaya and its significance in the collision tectonics, Active fault research 22 (memorial issue of Professor Tokihiko Matsuda) 16 July 2002

Pathak V, Pant CC, Darmwal GS (2015) Geomorphological features of active tectonics and ongoing seismicity of northeastern Kumaun Himalaya, Uttarakhand India. J Earth Syst Sci 124 (6):1143–1157

Pandey AK, Pandey P, Singh GD, Juya N (2014) Climate footprints in the late quaternary holocene landforms of Dun valley, NW Himalaya, India. Curr Sci 106:245–253

Philip G (2007) Remote sensing data analysis for mapping active faults in the northwestern part of Kangra Valley, NW Himalaya, India. Int J Remote Sens 28(21):4745–4761

Wesnousky SG, Kumar S, Mohindra R, Thakur VC (1999) Uplift and convergence along the Himalayan Frontal Thrust. Tectonics 18(6):967–976

Yeats RS, Thakur VC (1998) Reassessment of earthquake hazard based on a fault-bend fold model of the Himalayan plate-boundary fault. Curr Sci 74:230–233

Chapter 3
Active Faults of the Kumaun and Garhwal Himalaya

3.1 Active Faults in the Dehradun Valley Between the Yamuna and Ganga Rivers

In the Sub-Himalayan zone between the Yamuna and Ganga rivers, rocks of the Siwalik Group are folded from south to north into the Mohand anticline in the frontal Siwalik Range, broad Dun syncline occupying the Dun Valley, and the overturned Santaurgarh anticline in northern part of the Dun Valley (Thakur et al. 2007). The southern margin of the Mohand anticline is marked by a physiographic and tectonic break between the range front and the alluvial plain defining the HFT. The northern margin of Dun, lying at the base of the Mussoorie Range, is marked by the MBT, which marks a major tectonic boundary between the pre-Tertiary Lesser Himalayan formations and Cenozoic sediments of the Sub-Himalaya. At some localities near Rajpur and Rispana Rao in the Dun Valley, the MBT dipping NE 40° brings the Krol Group Chandpur phyllite over the late Quaternary–Holocene Dun gravels, indicating Holocene to Recent tectonic activity along the thrust (Thakur 2013). In his geomorphological mapping of Dehradun, Nakata (1972) recognized three late Quaternary Dun surfaces, viz., Higher, Middle, and Lower Dun surfaces, and three fluvial terrace levels based on their elevation, degree of dissection, and soil development (Fig. 3.1). Dehradun is flanked to the south by the Mohand Range, a part of frontal Siwalik Range which is made of the Mohand anticline by folding of the Siwalik Group strata. The anticline is developed as a fault bend/propagation fold over the HFT during late Quaternary (Thakur et al. 2007). The southern margin of the anticline is defined by a contact between the Siwalik strata and the recent alluvium. The trace of the contact showing a topographic break between the mountain front and the alluvial plain demarcates the HFT (Fig. 3.1).

We dated the geomorphic surfaces using Optically Stimulated Luminescence (OSL) method (Table 3.1). The top of the Nagsidh Hill representing the hill top surface yields an age of 46.3 ± 4.5 ka (sample NN-1). The Upper Dun surface is

© Springer Nature Singapore Pte Ltd. 2018
R. Jayangondaperumal et al., *Active Tectonics of Kumaun and Garhwal Himalaya*,
Springer Natural Hazards, https://doi.org/10.1007/978-981-10-8243-6_3

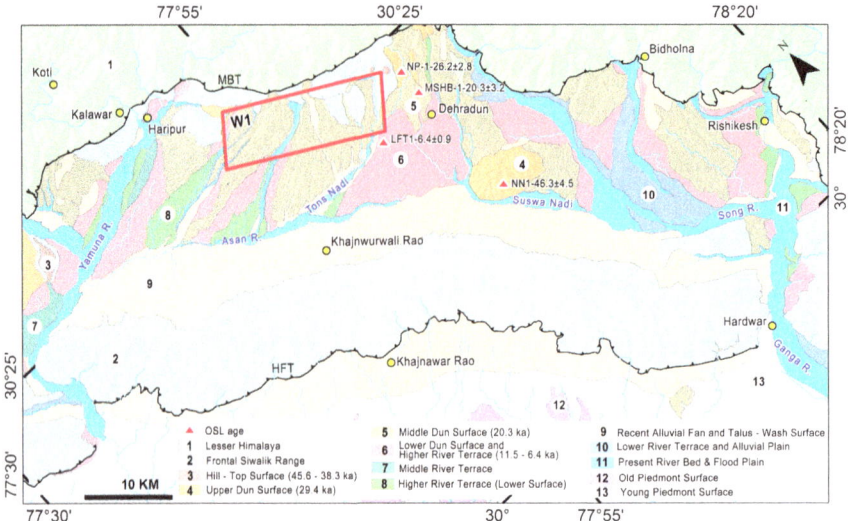

Fig. 3.1 Geomorphic map of Dehradun Valley shows geomorphic features including fan and terrace surfaces (modified after Nakata 1972). W1 is the window for active fault mapping undertaken in the present investigation. Red triangle indicates OSL age of the different surfaces. A separate table shows sample number along with OSL data (Table 3.1)

assigned an age of ∼26.2 ± 2.8 ka (NP-1). The Middle Dun surface exposed at the Hathibarkala estate yields an age estimate of 20.3 ± 3.2 ka (MSHB-1). The Lower Dun surface covering the suburb region of southern Dehradun has been assigned an age of 6.4 ± 0.9 ka (LFT1). A shortening of ∼6–16 mm/year has been estimated across the Dehradun reentrant based on a balanced cross section (Power et al. 1998). Based on the dating of strath terraces on the range front, a shortening rate of 11.9 ± 3.1 mm/year on the HFT has been estimated (Wesnousky et al. 1999).

3.1.1 Santaurgarh Thrust

The Window-1 represents the area demarcated in Fig. 3.1, where active faults are studied in detail (Figs. 3.2, 3.3, 3.4, 3.5, 3.6, 3.7, 3.8, 3.9, and 3.10). Two active faults, the Santaurgarh Thrust and Bhauwala Thrust (BT) and a smaller fault in between the two were mapped (Fig. 3.2). The faults affect the upper and Middle Dun surfaces forming scarps with vertical offsets. The fault traces of Santaurgarh and BTs show break in slope and offset of the Dun fan surfaces (Fig. 3.4). South of the MBT at several localities, the higher topography of the dissected Siwaliks abuts against the Dun gravels over the pedimented Siwaliks with abrupt break in slope along Santaurgarh Thrust (Fig. 3.3). The knee-bend deflection of rivers coincides with the thrust contact (Thakur et al. 2007).

Table 3.1 OSL age of late Quaternary deposits from Dehradun principal fans and Nagsidh Hill

S. No.	Sample No.	U (ppm)	Th (ppm)	Potassium K (%)	Dry water content (%)	Dose rate (Gy/ka)	Equivalent dose (Gy)	Age (ka)
1	NP-1	4.9 ± 0.1	21.6 ± 0.2	2.8 ± 0.0	9.47	130.5 ± 13.6	5.0 ± 0.1	26.2 ± 2.8
2	NN1	3.1 ± 0.3	13.4 ± 1.3	2.22 ± 0.2	16.78 ± 3.4	3.4 ± 0.3	158.0 ± 9.5	46.3 ± 4.5
3	LFT1	4 ± 0.4	15.9 ± 1.6	3.13 ± 0.3	14.63 ± 2.8	4.6 ± 0.4	29.6 ± 3.4	6.4 ± 0.9
4	MSHB-1	2.5 ± 0.3	10.5 ± 1.1	2.42 ± 0.2	2.42 ± 0.5	4.0 ± 0.5	80.4 ± 7.9	20.3 ± 3.2

Table 3.2 Characteristics of the Dehradun Valley fault scarps

Ref. No	Thrust	Location	Latitude (N)	Longitude (E)	Strike direction	Fault age	Length (m)	Fault feature	Fault reference
1	Santaurgarh Thrust	Northeast of Dungakhet	30.46367	77.9002	NW–SE	Post 500 ka	223	Fault scarp	Proximal—north part of Donga Fan surface
2	Santaurgarh Thrust	Northeast of Dungakhet	30.45874	77.9069	NNW–SSE	Post 500 ka	218	Fault scarp	Proximal—north part of Donga Fan surface
3	Santaurgarh Thrust	Southwest of Barwa	30.45242	77.90812	NNW–SSE	Post 500 ka	758	Fault scarp	Proximal—north part of Donga Fan surface
4	Santaurgarh Thrust	Kotra	30.43837	77.92501	NW–SE	Post 500 ka	1373	Fault scarp	Proximal—north part of Donga Fan surface
5	Santaurgarh Thrust	South of Birsani	30.43181	77.93078	NW–SE	Post 500 ka	565	Inferred fault scarp	Proximal—north part of Donga Fan surface
6	Santaurgarh Thrust	South of Dudhai	30.42713	77.93718	NWW–SEE	Post 500 ka	499	Fault scarp	Proximal—north part of Donga Fan surface
7	Santaurgarh Thrust	Southeast of Dudhai	30.42641	77.94068	NWW–SEE	Post 500 ka	88	Fault scarp	Proximal—north part of Donga Fan surface
8	Santaurgarh Thrust	North of Donga block	30.41867	77.95611	NW–SE	Post 500 ka	721	Inferred fault scarp	Proximal—north part of Donga Fan surface
9	Santaurgarh Thrust	Northwest of Bidhauli	30.41446	77.96899	NWW–SEE	Post 500 ka	659	Inferred fault scarp	Proximal—north part of Donga Fan surface
10	Santaurgarh Thrust	Badiwala	30.40932	77.98759	NWW–SEE	Post 500 ka	332	Inferred fault scarp	Proximal—north part of Donga Fan surface
11	Majhaun Thrust	North of Rajauli	30.43698	77.90336	NW–SE	–	294	North-facing fault scarp (back thrust)	Proximal—middle part of Donga Fan surface
12	Majhaun Thrust	Northeast of Rajauli	30.42775	77.9095	NW–SE	–	274	North-facing fault scarp (back thrust)	Proximal—middle part of Donga Fan surface

(continued)

Table 3.2 (continued)

Ref. No	Thrust	Location	Latitude (N)	Longitude (E)	Strike direction	Fault age	Length (m)	Fault feature	Fault reference
13	Majhaun Thrust	Northeast of Rajauli	30.41925	77.92206	NW–SE	–	724	North-facing fault scarp (back thrust)	Proximal—middle part of Donga Fan surface
14	Bhauwala Thrust	Southwest of Dungakhet	30.4602	77.88634	NNW–SSE	29–20 ka	433	Fault scarp	Proximal—middle part of Donga Fan surface
15	Bhauwala Thrust	South of Dungakhet	30.45254	77.88752	NNW–SSE	29–20 ka	434	Fault scarp	Proximal—middle part of Donga Fan surface
16	Bhauwala Thrust	Southwest of Kumhar Matti	30.44612	77.88935	NNW–SSE	29–20 ka	520	Fault scarp	Proximal—middle part of Donga Fan surface
17	Bhauwala Thrust	Southeast of Kumhar Matti	30.43858	77.89295	N–S	29–20 ka	216	Fault scarp	Proximal—middle part of Donga Fan surface
18	Bhauwala Thrust	Southeast of Kumhar Matti	30.43464	77.88804	NNW–SSE	29–20 ka	274	Fault scarp	Proximal—middle part of Donga Fan surface
19	Bhauwala Thrust	South of Rajauli	30.42593	77.89583	NW–SE	29–20 ka	1035	Inferred fault scarp	Proximal—middle part of Donga Fan surface
20	Bhauwala Thrust	Southeast of Rajauli	30.41996	77.90209	NW–SE	29–20 ka	338	Fault scarp	Proximal—middle part of Donga Fan surface
21	Bhauwala Thrust	Tilwari	30.41487	77.90775	NW–SE	29–20 ka	689	Inferred fault scarp	Proximal—middle part of Donga Fan surface
22	Bhauwala Thrust	East of Tilwari	30.41088	77.91251	NW–SE	29–20 ka	246	Inferred fault scarp	Proximal—middle part of Donga Fan surface
23	Bhauwala Thrust	Southwest of Donga block	30.40091	77.93309	NW–SE	29–20 ka	1379	Fault scarp	Proximal—middle part of Donga Fan surface
24	Bhauwala Thrust	North of Manduwala	30.39411	77.94373	NW–SE	29–20 ka	688	Fault scarp	Proximal—middle part of Donga Fan surface

(continued)

Table 3.2 (continued)

Ref. No	Thrust	Location	Latitude (N)	Longitude (E)	Strike direction	Fault age	Length (m)	Fault feature	Fault reference
25	Bhauwala Thrust	Northwest of Kanswali	30.38671	77.95646	NW–SE	29–15 ka	386	Fault scarp	Proximal—middle part of Donga Fan surface
26	Bhauwala Thrust	Northeast of Kanswali	30.38248	77.96654	NW–SE	29–15 ka	1198	Fault scarp	Proximal—middle part of Donga Fan surface
27	Bhauwala Thrust	Northwest of Paundha	30.38099	77.97556	NWW–SEE	29–20 ka	248	Fault scarp	Proximal—middle part of Donga Fan surface
28	Bhauwala Thrust	Northeast of Paundha	30.37793	77.98704	NW–SE	29–20 ka	339	Fault scarp	Proximal—middle part of Donga Fan surface
29	Bhauwala Thrust	Northeast of Paundha	30.36972	77.9965	NW–SE	29–20 ka	1228	Fault scarp	Proximal—middle part of Donga Fan surface

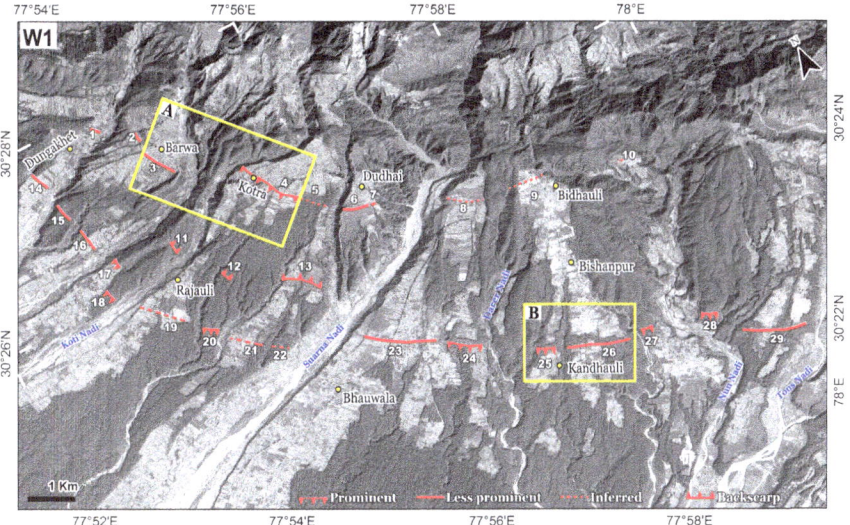

Fig. 3.2 Cartosat-1A grayscale satellite image showing fault characteristics of the W1 window (Fig. 3.1). Solid, comb, and dotted lines indicate active faults in the Dehradun Valley. Numerals adjoining the active faults (shown in red color) denote the characteristics of the fault given in Table 3.4. Contour map and geomorphic map of the W1 window is given in Figs. 3.3 and 3.4. Rectangles A (Fig. 3.5) and B (Fig. 3.6) show detailed characteristics of active faults

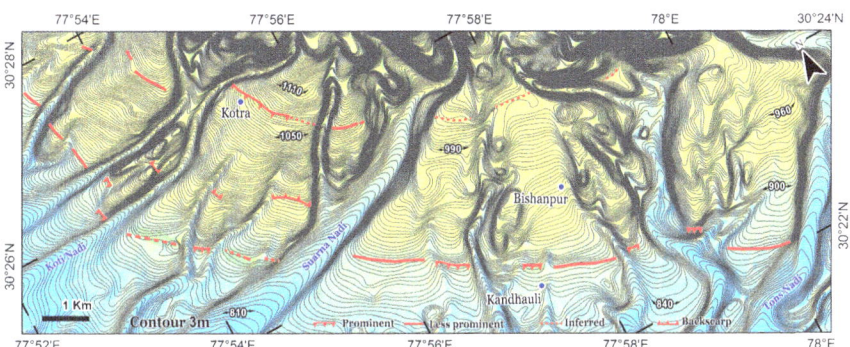

Fig. 3.3 Contour lines endorse fault scarps of the W1 window. The closely spaced contour lines indicating steepness of the slopes. Contour lines (3 m) are derived from Cartosat-1A digital terrain surface. Location of this figure is marked as a rectangle W1 in Fig. 3.1

3.1.2 Bhauwala Thrust

South of the Santaurgarh Thrust, the N–S trending isolated ridges of thick Dun gravels with pedimented Siwaliks base abruptly terminate with a topographic break. The southern tips of these ridges are aligned along a prominent lineament with

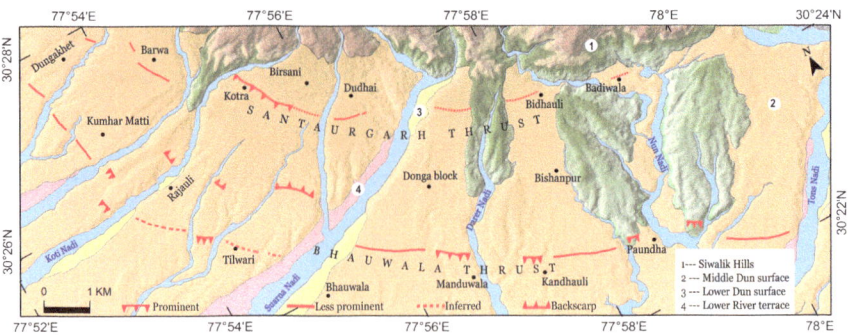

Fig. 3.4 Fault traces of the Santaurgarh and Bhauwala thrusts are characterized with break in slope and formation of scarps within the Dun fan surfaces between Dungakhet and Tons Nadi. Location of this figure is marked as a rectangle W1 in Fig. 3.1

Fig. 3.5 Three-dimensional perspective view of sub-window A (Fig. 3.2) toward northeast direction shows prominent faults at Barwa and Kotra (red arrow). Cartosat-1A grayscale imagery draped over 10 m digital terrain surface. Black solid line indicates location of profile cross section (P1) (Fig. 3.6). Field photograph location is marked as Fp1 (Fig. 3.7). Numerals indicate fault reference given in Fig. 3.2 and Table 3.2

uplifted northern block representing the BT. The BT marks the southern boundary of the steeply dipping pedimented Siwaliks with the Dun gravel cover. This geomorphological and tectonic expression has been observed along all the river sections in the Donga Fan. The BT continues further eastward in the Dun Principal fan, where streams make a sharp knee-bend turn along the lineament. On the basis of the OSL ages of the Dun gravels and their relative positions, the timing of fault activity

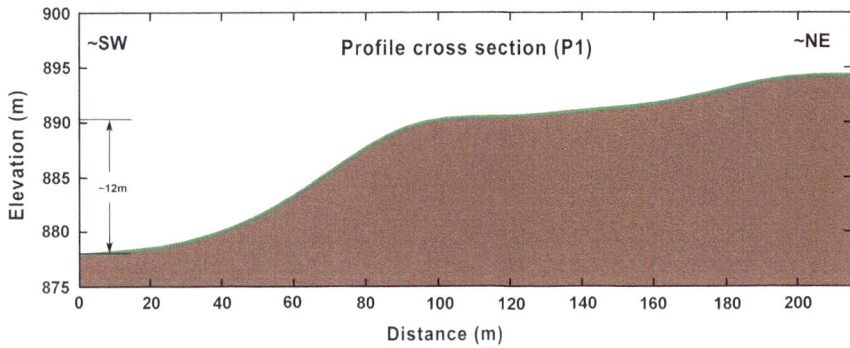

Fig. 3.6 Profile across the fault scarp in northern part of Middle dun surface around Kotra area. Location of P1 is marked in Fig. 3.5

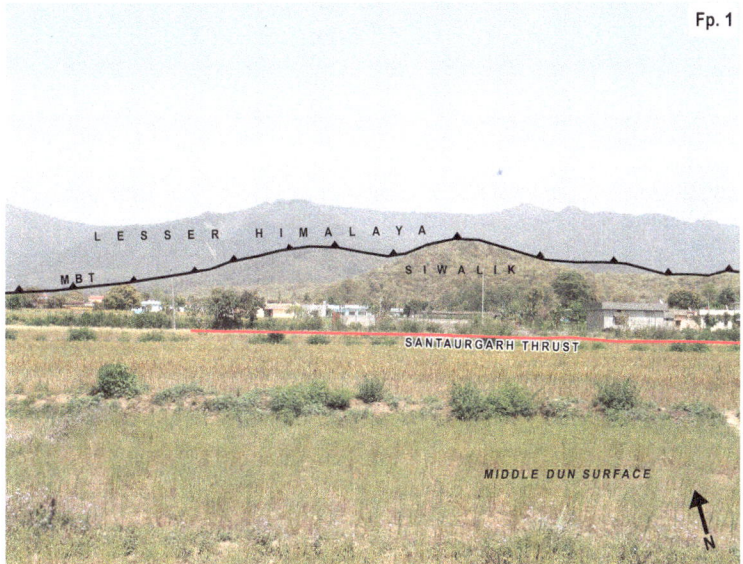

Fig. 3.7 Field photograph shows MBT, Santaurgarh Thrust, and Middle dun surface. Location of Fp1 is marked in Fig. 3.5

has been evaluated (Thakur et al. 2007). To the north of the BT in the proximal part of the Donga Fan, the Unit-B gravels dated 29.4 ± 1.7 ka by OSL lie at ∼ 800 m elevation, whereas to the south of the thrust 29.5 ± 5.0 ka dated gravels lie at ∼ 700 m elevation in the middle part of the fan surface.

This difference in the elevation of a fan surface with the same age on both sides of the BT indicates faulting activity post ∼ 29 ka. The younger fan surface with Unit-C gravel ages between 22.8 and 10.7 ka is not disrupted by the BT, defining the lower age limit activity along the thrust. Near the Kandhauli Village, ∼ 10 m

Fig. 3.8 Three-dimensional perspective view of sub-window B (Fig. 3.2) toward northern direction shows prominent fault near Kandhauli (red arrow). Black solid line indicates location of profile cross section (P2) (Fig. 3.9). Green triangle is landmark of trench site along the profile P2 (Fig. 3.10). Field photograph locations are marked as Fp2 and Fp3 (Fig. 3.10). Numerals indicate fault reference given in Fig. 3.2 and Table 3.2

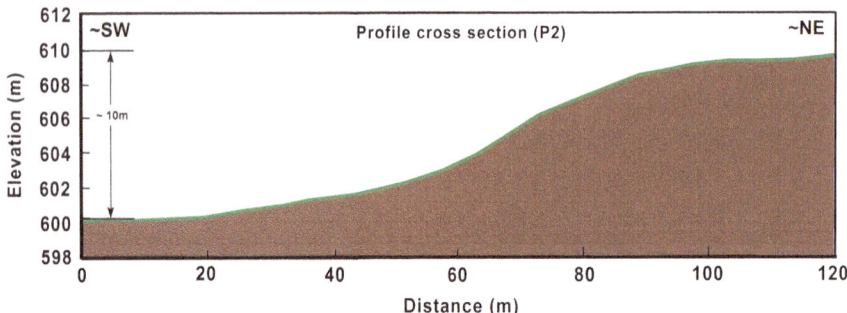

Fig. 3.9 Cross-profile variation of the prominent fault scarps. Location of P2 is marked in Fig. 3.8. Trench log is given in Fig. 3.10

high tectonic scarp of the BT is exposed (Fig. 3.9). The scarp trends NW–SE and formed within the Dun gravels. The lithological layering in paleoseismic trench wall shows a monocline fold. No fault is located in the ~5 m deep trench, implying the scarp is a fold scarp formed as a result of fault propagation of a blind fault that may lie at a greater depth than the excavated trench. OSL dating of the Dun gravels exposed in the trench wall indicates the formation of fold scarp after ~15.0 ka (Fig. 3.10).

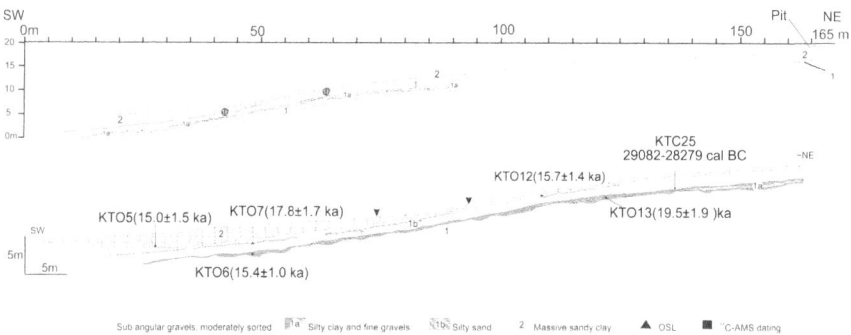

Fig. 3.10 (Top) Trench log of Bhauwala fault scarp, (bottom) trench exposure profile with pit on the hanging wall

In the Dun surface, the Santaurgarh and BT are identified in the geomorphic map as well as orthorectified Cartosat-1A satellite images (Figs. 3.2 and 3.4). Fault scarps of the Santaurgarh and BTs and the third small fault are aligned along the closely spaced intervals of contour lines indicative of steep slopes (Fig. 3.3). In Fig. 3.2, yellow rectangles represent sub-windows A and B. Numerals in this figure refer to the fault characteristics given in Table 3.3.

In the sub-window A, the western portion of the Santaurgarh Thrust exposed in Barwa and Kotra villages is marked in the Cartosat-1A overlaying the digital elevation map (Fig. 3.5). Bar P1 in the figure shows the location of the topographic profile. The topographic profile constructed along the P1 shows ∼12 m height of the fault scarp indicating vertical offset (Fig. 3.6). A type example of the field photograph shows the MBT, Santaurgarh Thrust, and Middle dun surface (Fig. 3.7). The sub-window B in yellow rectangle (Fig. 3.2) denotes the BT mapped in the Kandhauli Village. Near Kandhauli, three-dimensional prospective view constructed across the BT trace shows well-defined fault scarp (Fig. 3.8). Bar P2 is the location of the elevation profile across the fault scarps constructed. The cross profile shows ∼10 m height of the fault scarps indicating fault offsets (Fig. 3.9). A trench was excavated across the scarp, and illustrated log is provided in Fig. 3.10. A type example of the BT fault scarp mapped in the field is shown in Fig. 3.11.

Table 3.3 Showing OSL ages obtained from Kandhauli trench, Bhauwala Thrust

Sample No.	Laboratory No.	U (ppm)	Th (ppm)	Potassium K (%)	Water content (%)	Dose rate (Gy/ka)	Equivalent dose (Gy)	Age (ka)
KTO-5	LD390	3.1 ± 0.03	22.2 ± 0.22	2.40 ± 0.02	3.57	4.7 ± 0.1	70.5 ± 6.8	15.0 ± 1.5
KTO-7	LD391	2.8 ± 0.03	22.2 ± 0.22	2.80 ± 0.03	2.26	5.1 ± 0.1	90.9 ± 8.8	17.8 ± 1.7
KTO-12	LD392	3.6 ± 0.04	21.3 ± 0.21	2.32 ± 0.02	3.51	4.7 ± 0.1	73.8 ± 6.4	15.7 ± 1.4
KTO-13	LD393	5 ± 0.05	18.8 ± 0.19	2.54 ± 0.02	2.23	5.1 ± 0.1	100.1 ± 9.4	19.5 ± 1.9
KTO-6	LD395	2.9 ± 0.03	21.6 ± 0.22	2.60 ± 0.03	2.75	4.9 ± 0.1	75.3 ± 5.0	15.4 ± 1.0

Fig. 3.11 Active fault scarp (red dotted line) displacing medial part of Middle dun fan surface near Kandhauli. Locations of Fp2 and Fp3 are marked in Fig. 3.8

Table 3.4 Active faults along the HFT at the Mohand anticline

Ref. No	Thrust	Location	Latitude (N)	Longitude (E)	Strike	Fault age	Length (m)	Fault feature	Fault reference
1	HFT	West of Hindan River	30° 13' 21.03"N	77° 47' 16.75"E	NW–SE	–	142	Fault scarp	Higher river terrace—southern part of Mohand anticline
2	HFT	East of Solani River	30° 11' 55.50"N	77° 49' 53.42"E	NNW–SSE	–	343	Fault scarp	Higher river terrace—southern part of Mohand anticline
3	HFT	North of Pelo Khurd	30° 10' 56.15"N	77° 52' 37.62"E	NNW–SSE	–	373	Fault scarp	Higher river terrace—southern part of Mohand anticline
4	HFT	East of Mohand Rao	30° 10' 11.09"N	77° 54' 23.55"E	NW–SE	–	689	Fault scarp	Higher river terrace—southern part of Mohand anticline
5	HFT	East of Mohand Rao	30° 10' 01.43"N	77° 54' 48.82"E	NW–SE	–	619	Inferred fault scarp	Higher river terrace—southern part of Mohand anticline
6	HFT	East of Chilawal Rao	30° 08' 52.73"N	77° 56' 21.05"E	NWW–SEE	Post 3663 ± 215 BP	219	Fault scarp	Higher river terrace—southern part of Mohand anticline

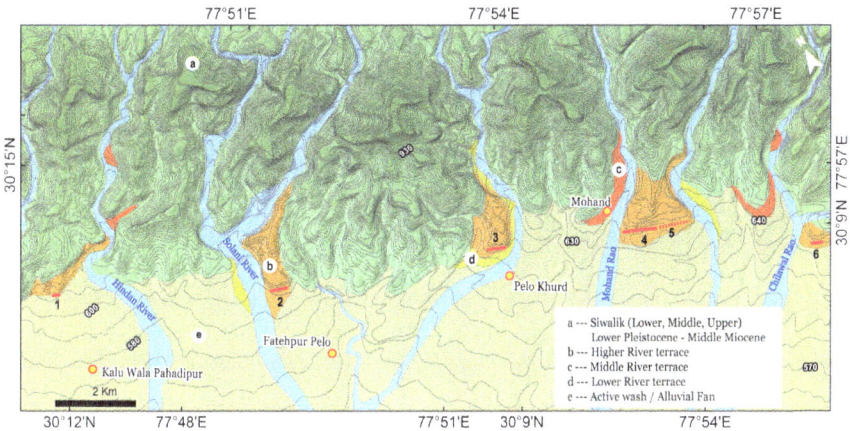

Fig. 3.12 Tectono-geomorphic and contour map of the Mohand anticline, Sub-Himalayan front (HFT). Scarps are aligned along the closely spaced contour lines indicating steepness of the slopes. Contour lines (10 m) are derived from Cartosat-1A digital terrain surface model

3.1.3   Active Faults Along the HFT

Tectono-geomorphic map has been made of the Sub-Himalayan front between the Hindan and Chilawal Rao rivers, southern part of Mohand anticline (Fig. 3.11). Axial trace of the Mohand anticline, HFT trace, and three levels of terraces have been mapped. Numerals in this figure refer to the fault characteristics given in Table 3.4. Based on the radiocarbon dating of detrital charcoal at the strath terrace (3663 ± 215 radiocarbon years B.P.) on the range front, a shortening rate of 11.9 ± 3.1 mm/year has been estimated on the HFT (Wesnousky et al. 1999). Contour lines derived from Cartosat-1A of the Sub-Himalayan front in the area show steepness of contour intervals indicating eroded fault scarps along the HFT (Figs. 3.11 and 3.12).

Anaglyph images for active faults along Dehradun valley between Yamuna and Ganga Rivers (Fault Reference number in the figure is corresponds to characteristics of the fault given in Table 3.2).

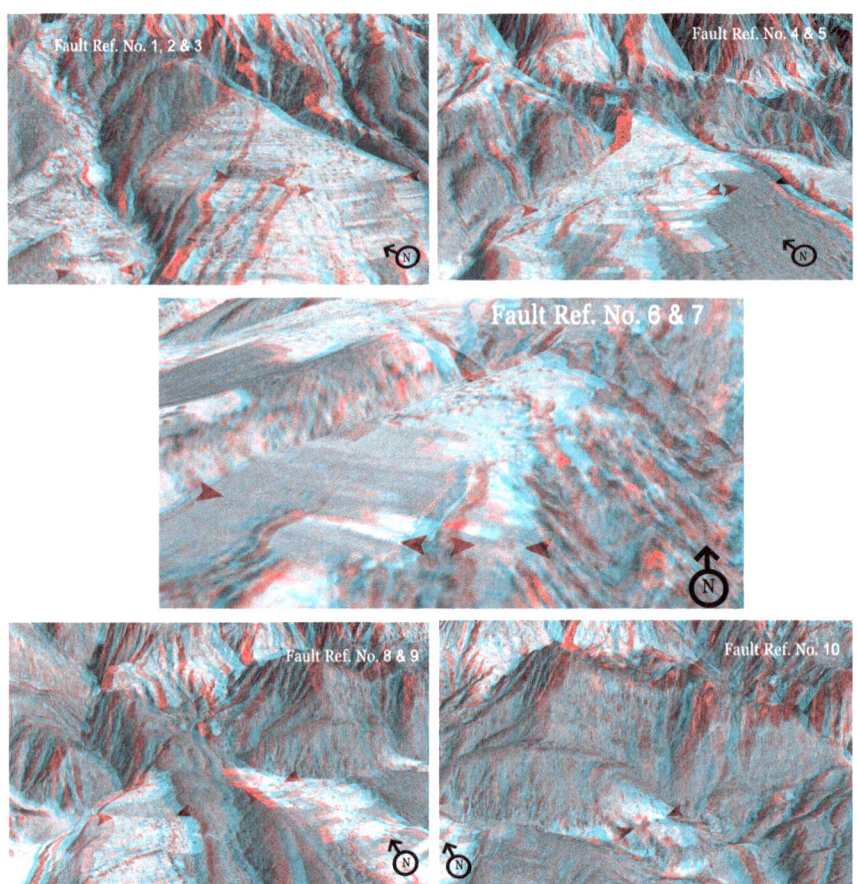

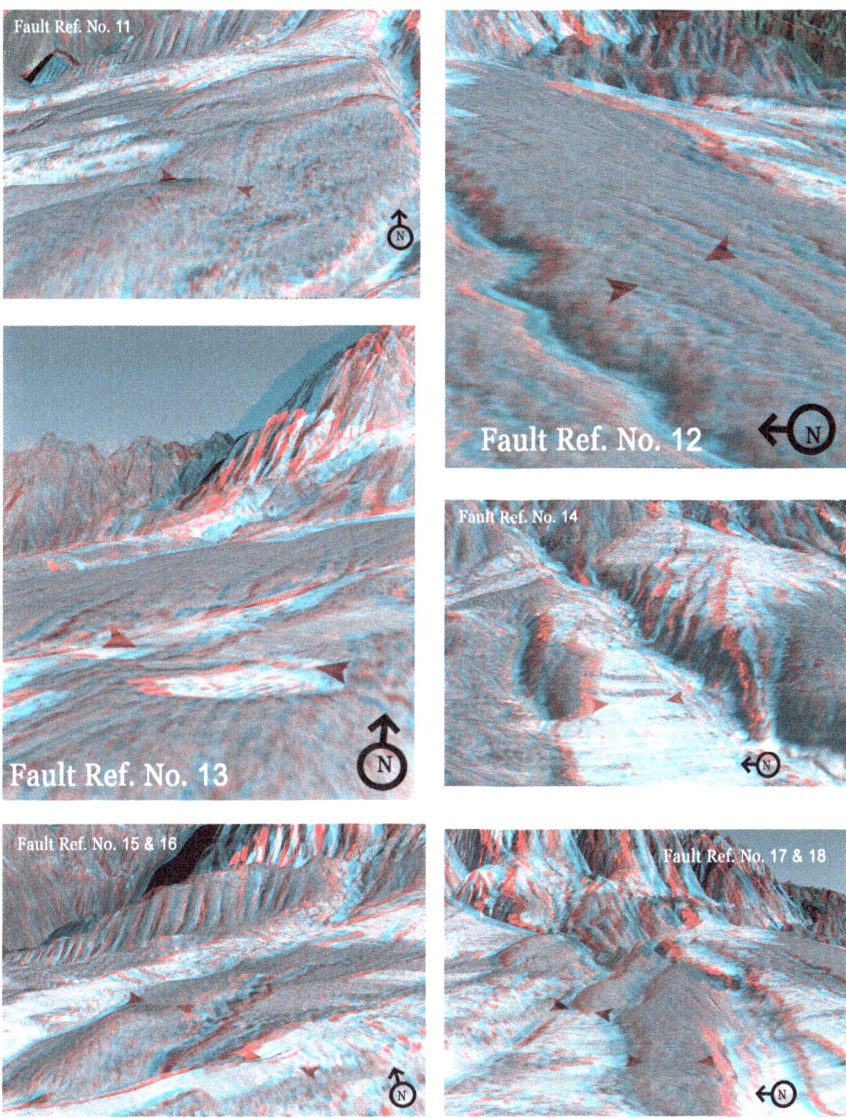

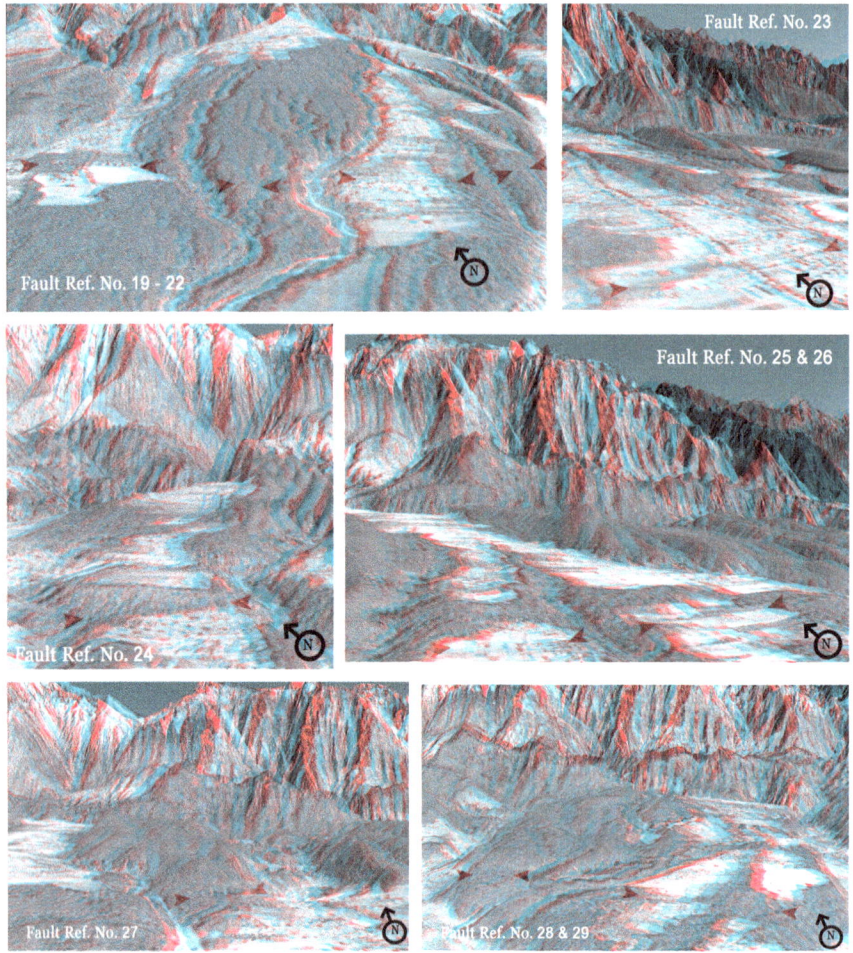

3.2   Active Faults Between Laldhang and Kotdwar

The fault trace of the HFT is recognized in near-vertical scarps along the Sub-Himalayan front at localities Laldhang, west of Trilokpur, and north of Padampur. The HFT trace is marked as an abrupt physiographic break between the Sub-Himalayan front and the alluvial plain, which is well displayed in the Cartosat-1A stereopair image (Fig. 3.13). In Fig. 3.13, the numerals refer to fault characteristics mentioned in Table 3.5, and the rectangle A (sub-window) refers to the location of the excavated paleoseismological trench (Fig. 3.19) and detailed mapping. The relationship between the topographic slope and the contour interval

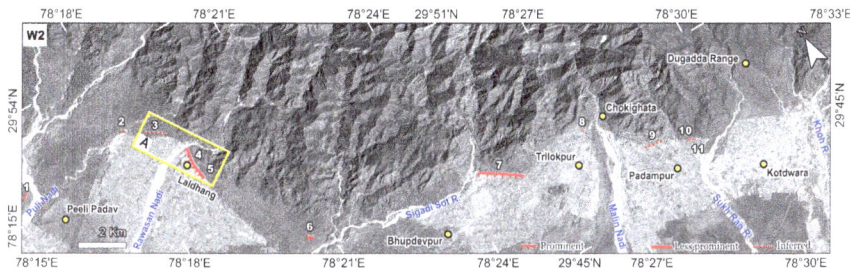

Fig. 3.13 Cartosat-1A satellite imagery showing fault scarps between the Puli Nadi and the Khoh River. Numerals adjoining active fault (shown in red color) refer to characteristics of the fault given in Table 3.5. Contour map and geomorphic map of the W1 window is given in Figs. 3.14 and 3.15. Rectangle A shows detailed characteristics of active fault (Fig. 3.16). Profile cross section along the scarp is shown in Fig. 3.17. Trench excavation and details are shown in Figs. 3.19, 3.20, and 3.21

Table 3.5 Characteristics of the fault scarps in east of Ganga River between Laldhang and Kotdwar

Ref. No	Thrust	Location	Latitude (N)	Longitude (E)	Strike	Length (m)	Fault feature	Fault reference
1	HFT	North of Peeli Padav	29° 52' 20.50"N	78° 15' 24.98"E	NEE–SWW	307	Fault scarp	Terrace
2	HFT	Northeast of Peeli Padav	29° 52' 28.13"N	78° 18' 02.13"E	NW–SE	274	Fault scarp	Surface
3	HFT	North of Laldhang	29° 52' 03.78"N	78° 18' 39.65"E	NW–SE	967	Fault scarp	Surface
4	HFT	Laldhang	29° 51' 10.80"N	78° 19' 04.50"E	NNW–SSE	532	Fault scarp	Terrace
5	HFT	South of Laldhang	29° 50' 50.43"N	78° 19' 05.55"E	NNW–SSE	686	Fault scarp	Surface
6	HFT	Northwest of Bhupdevpur	29° 48' 19.94"N	78° 20' 32.81"E	NW–SE	260	Fault scarp	Fan distal part
7	HFT	West of Trilokpur	29° 47' 23.54"N	78° 24' 55.75"E	NW–SE	1776	Fault scarp	Surface and terrace
8	HFT	North of Trilokpur	29° 47' 18.79"N	78° 27' 00.64"E	NW–SE	95	Fault scarp	Terrace
9	HFT	Northwest of Padampur	29° 46' 15.39"N	78° 28' 14.43"E	NWW–SEE	776	Fault scarp	Surface
10	HFT	North of Kotdwar	29° 45' 56.92"N	78° 28' 58.64"E	NW–SE	142	Fault scarp	Surface
11	HFT	North of Kotdwar	29° 45' 53.97"N	78° 29' 05.17"E	NW–SE	125	Fault scarp	Surface

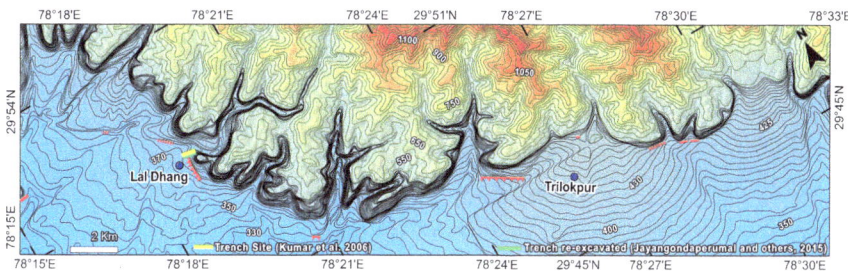

Fig. 3.14 Fault scarps are aligned along the closely spaced contour lines indicating steepness of the slopes. Contour lines in fan surface show 5 m interval (black tone), and contour lines in Siwalik Formation show 50 m interval (dark green tone). Contour lines are derived from Cartosat-1A digital terrain surface. Yellow rectangle shows trench location (Kumar et al. 2006) across the fault scarp

Fig. 3.15 Fault traces of Himalayan Frontal Thrust showing scarps and offsets of the alluvial fan surfaces along the Himalayan front

shows that the fault scarps lie along the steepest gradient (Fig. 3.14). The fault scarps displace the contemporary geomorphic surfaces (Fig. 3.15). Near Laldhang, an active fault scarp was mapped (sub-window A) in detail and shown in Cartosat-1A overlaying DEM data (Fig. 3.16). The elevation profile constructed normal to the HFT scarp shows ~10 m height (vertical offset) (Fig. 3.17). The field photographic view of the Laldhang area shows prominent linear NW–SE trending fault scarp of the HFT (Fig. 3.18).

3.2.1 Laldhang Trench

The HFT is exposed showing ~9 m high fault scarp in the Quaternary alluvium near the village of Laldhang, located about 20 km east of Haridwar. Kumar et al. (2006) made paleoseismic investigation, excavating ~25 m long and 4–6 m deep trench perpendicular to a northwest trending scarp (Fig. 3.19). The eastern wall of the trench was cleaned and logged on a 1 m × 1 m square grid to work out the trench wall stratigraphy. Seven stratigraphic units, 1–7, were recognized on the

Fig. 3.16 Detailed active fault trace (red line) in Laldhang village, regional location shown as rectangle A in Fig. 3.13. Thick black line P1 showing location of topographic profile constructed across the active fault is indicated in Fig. 3.17. Fp.1 is the field photographic location of Fig. 3.18

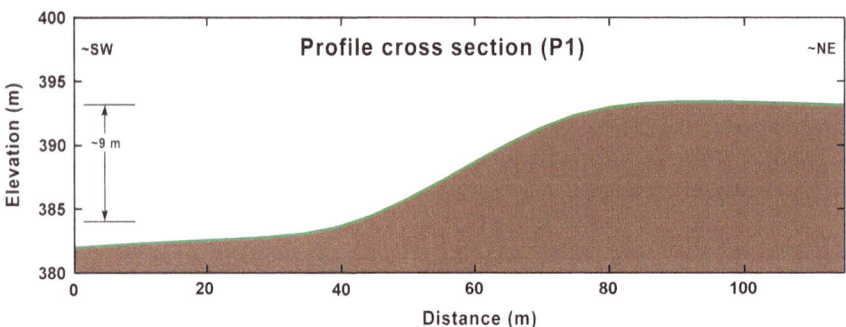

Fig. 3.17 Elevation profile across the prominent fault scarp. Profile location is marked in Fig. 3.16. Scarp height is ∼9 m corresponds to fault offset in the contemporary geomorphic surface

trench exposure, whose stratigraphic logs units are shown in Fig. 3.19. The unit-1 of alluvial fan gravels at the base is conformably overlain by units 2–5, constituting the footwall. The units 1–5 are repeated above the fault strand F1. The repeated units 1–5 are deformed above fault strand F1 and form the hanging wall. Unit-6 is not deformed, which implies that the top of unit-5 was the ground surface at the time of displacement along the fault strand 'F1'. The repetition of units 1–5 resulted due to deformation on the hanging wall as a result of slip along the horizontal fault

Fig. 3.18 Active fault scarp (red dotted line) displacing fan surface near Laldhang

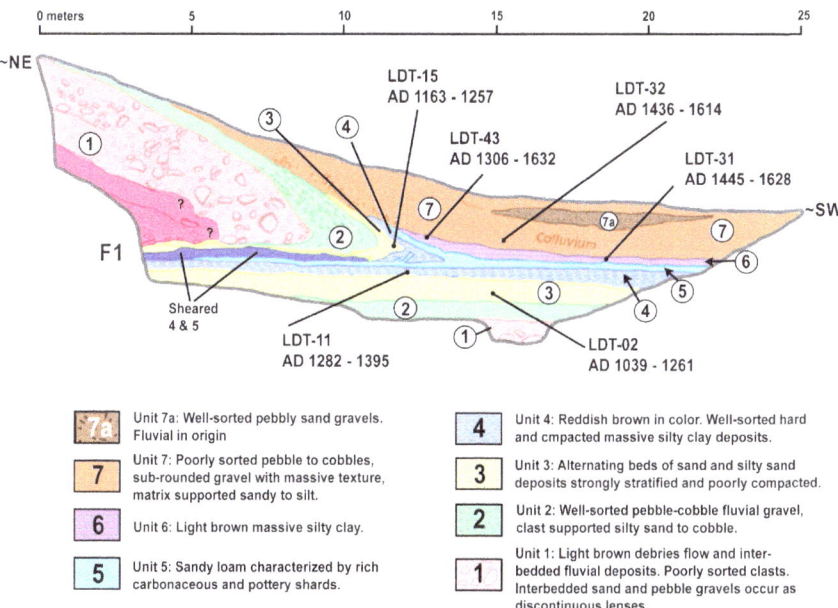

Fig. 3.19 Trench log of the Laldhang fault scarp (Kumar et al. 2006). Radiocarbon sample collection locations are shown in the trench log (modified from Jayangondaperumal et al. 2017b). The location of this figure is marked as T1 (star with yellow color) in Fig. 3.16

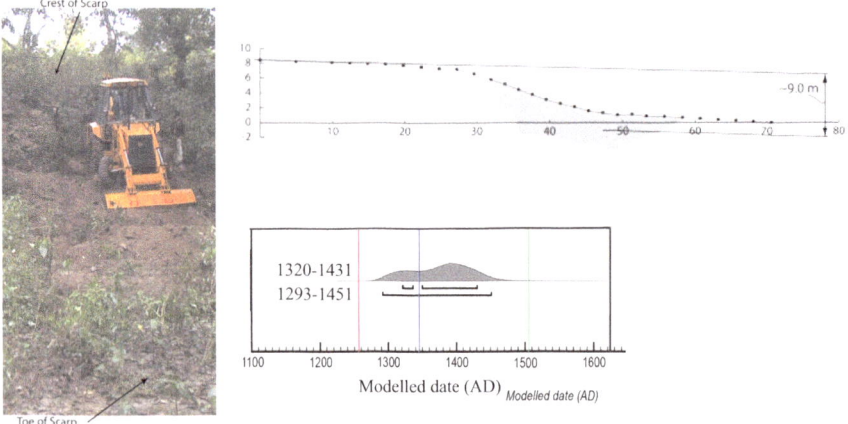

Fig. 3.20 Left: Field photograph of the Laldhang scarp (Kumar et al. 2006). Right top: Laldhang trench fault scarp elevation profile (Kumar et al. 2006). Right bottom: One- and two-sigma confidence intervals are given for rupture dates at Laldhang site, alongside probability distribution functions developed from radiocarbon age results using the OxCal software program (Bronk Ramsey 2009a, b). Vertical lines indicating dates of putative earthquakes in 1255 CE (red vertical line), 1344 CE (blue vertical line), and 1505 CE (green vertical line) are superimposed. The location of this trench is marked as T1 (star with yellow color) in Fig. 3.16

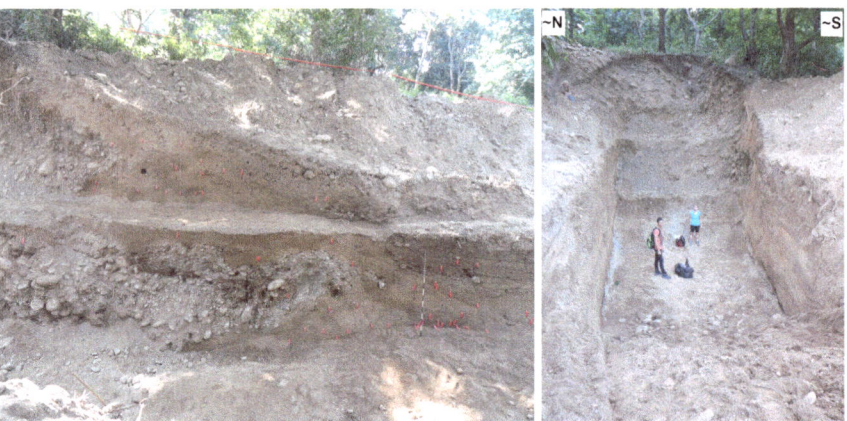

Fig. 3.21 Reexcavated trench wall exposure at Laldhang (Jayangondaperumal et al. 2017b). The location of this figure is marked as T2 (star with purple color) in Fig. 3.16

strand F1. Detrital charcoal samples obtained from unit-3 give dates that range between A.D. 1039 and 1261 (sample LDT-11) and obtained from unit-4 provides a maximum date of A.D. 1282–1395 for the fault displacement. Units 6 and 7 cap the deformed units 1–5, indicating that they were deposited post F1 faulting.

Two radiometric ages of A.D. 1306 and 1632 from detrital charcoal samples obtained from the unfaulted unit-6 (i.e., capping unit) postdate the fault displacement. The deformation features observed on the trench wall exposure and the radiocarbon ages suggest displacement occurred between A.D. 1282 and 1632 as a result of a single great earthquake (Fig. 3.19). More recently, modeling of published age data of Laldhang trench (Kumar et al. 2006) using Bayesian statistical program OxCal indicates an event, coeval on the HFT, which corresponds to a historically documented A.D. 1344 earthquake reported from Nepal (Pant 2002) with an estimated magnitude of >8.6 MW (Jayangondaperumal et al. 2017).

Anaglyph images for Laldhang and Kotdwar (Fault Reference number in the figure is corresponds to characteristics of the fault given in Table 3.5).

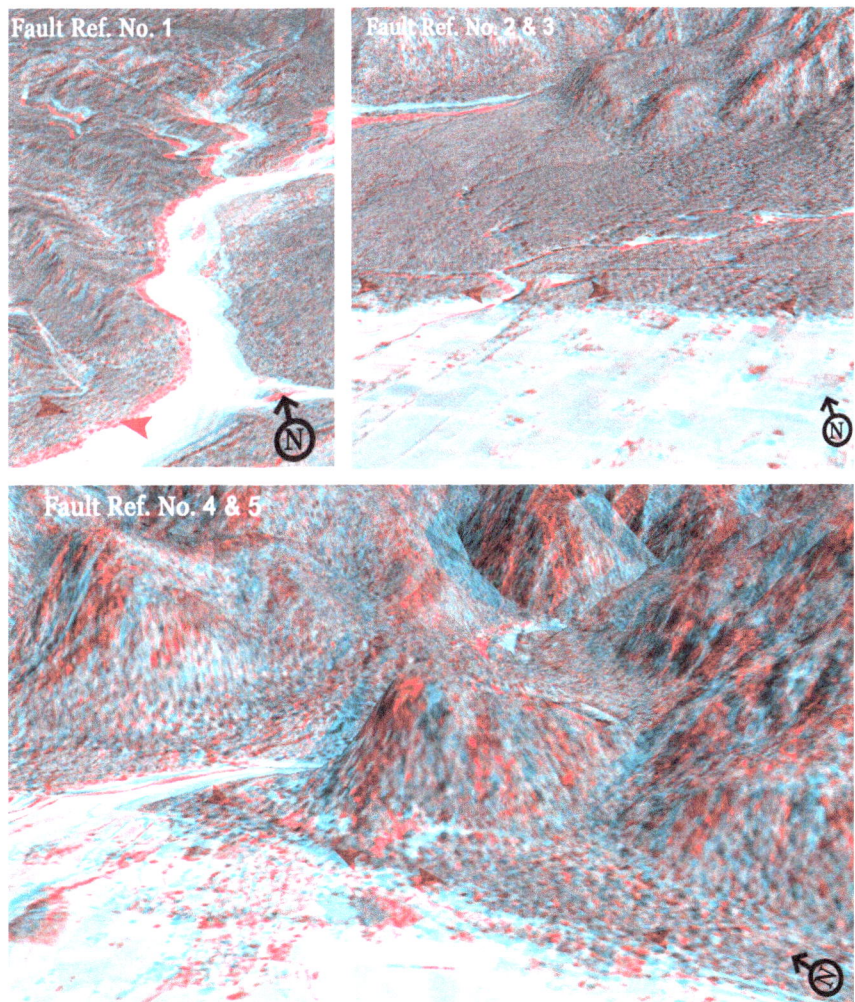

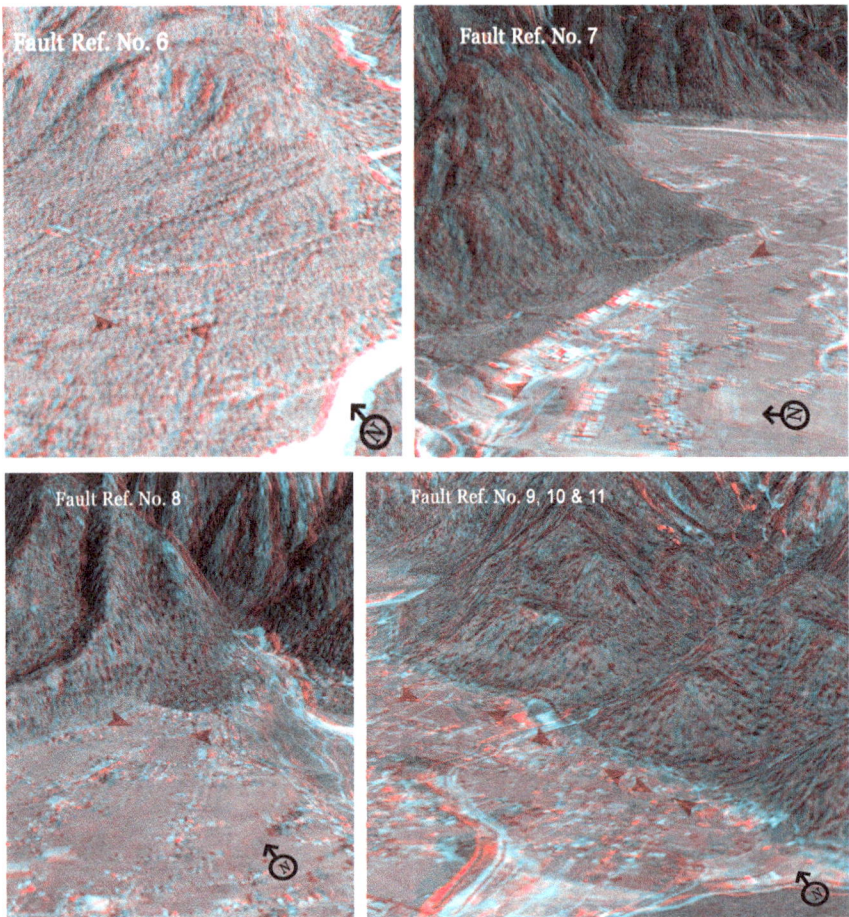

3.3 Active Faults in Kotadun

Nakata (1972) was the pioneer in the preparation of geomorphic maps of the Kotadun and identification of the HFT as the 'Himalayan Front Tectonic Line.' He recognized five levels of geomorphic surfaces and four branches of active faults. Later, active faults in the region were studied by Valdiya (1992, 2003). Kotadun, a small intermontane basin filled with the Dun gravels lies within the Siwaliks in central Kumaun Sub-Himalaya. Six differently named dun gavels fans and five levels of terraces have been mapped. The dun extends in NW–SE direction for a

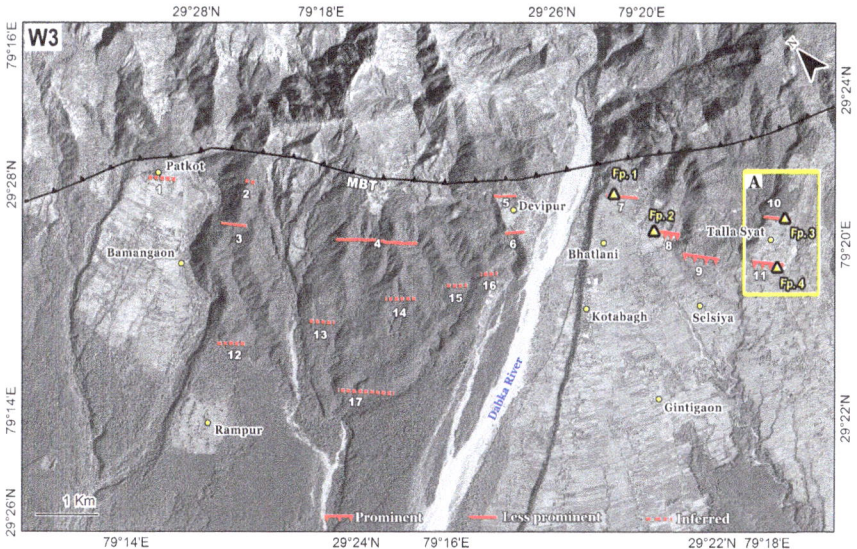

Fig. 3.22 Grayscale satellite image shows active faults between Palkot Sot and Baur river terrace. Numerals (1–17) adjoining active fault (shown in red color) correspond to characteristics of the fault given in Table 3.6. The same in contour and geomorphic map is given in Figs. 3.23 and 3.24. Rectangle A shows detailed characteristics of active faults (Fig. 3.25). Triangle Fp.1 to Fp.4 (shown in yellow color) indicates field photograph location of Fig. 3.28

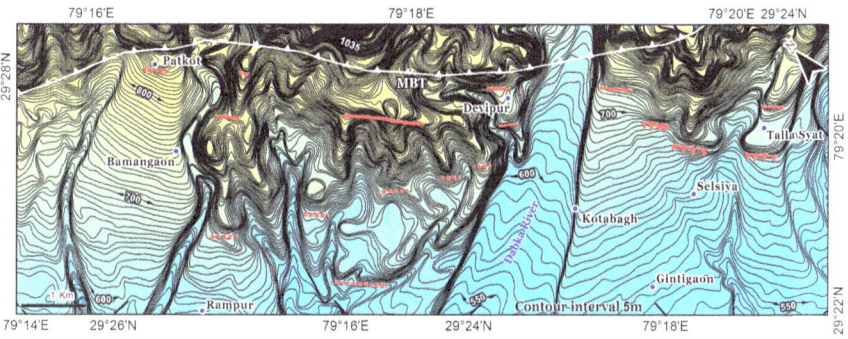

Fig. 3.23 Kotadun fault scarps are aligned along the closely spaced contour lines. Contour lines in 5 m interval are derived from Cartosat-IA DEM data

length of ∼20 km and a width of ∼5 km between the Kosi and Baur rivers. The Kotadun basin is bounded to the north by the Lower/Middle Siwalik Mountain along a fault called the Dhikala Thrust and to the south by the low lying hills of the Upper Siwaliks and in between the Pawalgarh Fault (Goswami and Pant 2007).

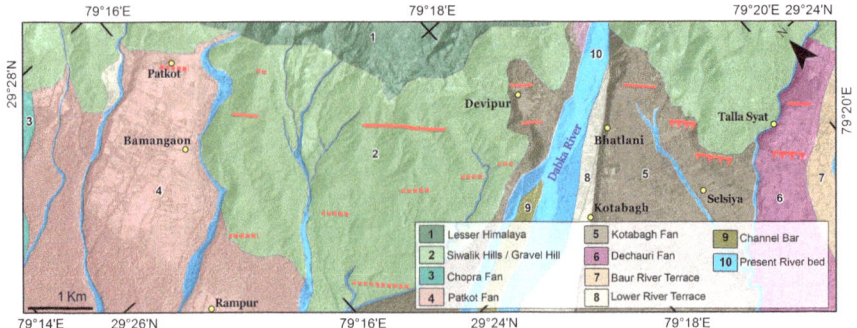

Fig. 3.24 Active fault traces show disjointing the alluvial fan surfaces and Siwalik/Gravel Hills between Palkot Sot and Baur River terrace

Fig. 3.25 A three-dimensional perspective view toward northeast direction shows two levels of prominent faults near Talla Syat (red arrow). Numerals are fault reference as shown in rectangle A of Fig. 3.22 and Table 3.6. Thick black line P1 showing topographic profile constructed across the active fault is indicated in Fig. 3.26. Cartosat-1A overlaid on 10 m Digital Elevation Model

The abrupt physiographic rise of the frontal Upper Siwalik Hills above the alluvial plain is demarcated at the base by the HFT. The Upper Siwalik strata overlain by the Dun Fan gravels have been uplifted as a consequence of displacement over the HFT suggesting late Quaternary initiation of the thrust fault.

Table 3.6 Characteristics of the Kotadun fault scarps

Ref. No	Thrust	Location	Latitude (° N)	Longitude (°E)	Strike	Length (m)	Fault feature	Fault reference
1	HFT	Northwest of Bamangaon	29.454304	79.269561	NW–SE	394	Fault scarp	Proximal part of Kosi Bagh Surface and Kosi middle river terrace
2	HFT	Northeast of Bamangaon	29.445625	79.278059	NW–SE	147	Fault scarp	Kosi higher river terrace
3	HFT	Northeast of Bamangaon	29.442781	79.272801	NW–SE	384	Fault scarp	Kosi higher river terrace
4	HFT	Southwest of Devipur	29.428347	79.286264	NW–SE	1162	Fault scarp	Proximal part of Kosi Bagh Surface and Kosi middle river terrace
5	HFT	North of Devipur	29.420127	79.303525	NW–SE	325	Fault scarp	Proximal part of Kosi Bagh Surface and Kosi middle river terrace
6	HFT	South of Devipur	29.415985	79.300565	NW–SE	285	Fault scarp	Proximal part of Kosi Bagh Surface and Kosi middle river terrace
7	HFT	North of Bhatlani	29.408751	79.316016	NW–SE	463	Fault scarp	Proximal part of Kosi Bagh Surface and Kosi middle river terrace
8	HFT	Northeast of Bhatlani	29.402146	79.316362	NW–SE	455	Fault scarp	Proximal part of Kosi Bagh Surface and Kosi middle river terrace
9	HFT	North of Selsiya	29.396349	79.317950	NW–SE	542	Fault scarp	Proximal part of Kosi Bagh Surface and Kosi middle river terrace
10	HFT	North of Talla Syat	29.392926	79.330025	NW–SE	336		Proximal part of Kosi Bagh Surface and Kosi middle river terrace
11	HFT	South of Talla Syat	29.389643	79.324114	NW–SE	492	Fault scarp	Proximal part of Kosi Bagh Surface and Kosi middle river terrace
12	HFT	North of Rampur	29.430907	79.261596	NW–SE	395	Fault scarp	Proximal part of Kosi Bagh Surface and Kosi middle river terrace

(continued)

Table 3.6 (continued)

Ref. No	Thrust	Location	Latitude (° N)	Longitude (°E)	Strike	Length (m)	Fault feature	Fault reference
13	HFT	Northeast of Rampur	29.425122	79.272443	NW–SE	348	Fault scarp	Proximal part of Kosi Bagh Surface and Kosi middle river terrace
14	HFT	Northeast of Rampur	29.424469	79.272882	NW–SE	446	Fault scarp	Proximal part of Kosi Bagh Surface and Kosi middle river terrace
15	HFT	West of Kotabagh	29.416228	79.289738	NW–SE	342	Fault scarp	Proximal part of Kosi Bagh Surface and Kosi middle river terrace
16	HFT	West of Kotabagh	29.414028	79.294179	NW–SE	248	Fault scarp	Proximal part of Kosi Bagh Surface and Kosi middle river terrace
17	HFT	Northeast of Rampur	29.414594	79.270156	NW–SE	806	Fault scarp	Proximal part of Kosi Bagh Surface and Kosi middle river terrace

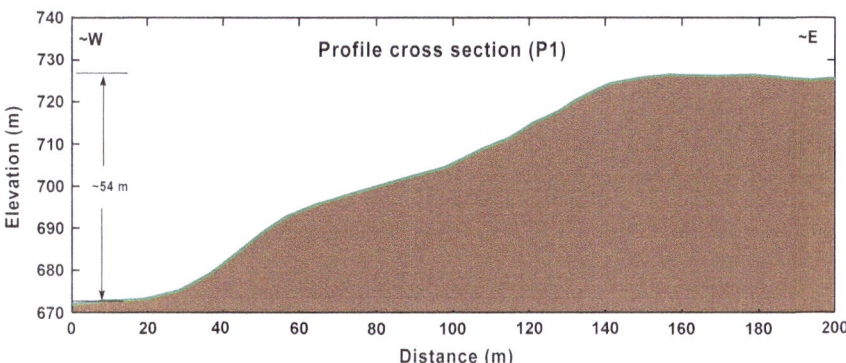

Fig. 3.26 Topographic profile across the prominent fault scarps. Location of profile cross section (P1) is marked in Fig. 3.25

Three traverses were made in the Kotadun, and the structures were reinterpreted. We focused on window-3 in the Kotadun for detailed active fault mapping (Fig. 3.22). Based on the analysis of Cartosat-1A image and field investigation, seventeen fault scarp segments (1–17) are identified in the window W-3 south of the MBT (Figs. 3.22, 3.23, 3.24, and 3.25). Numerals 1–17 in Fig. 3.22 refer to the fault characteristics given in Table 3.6. The scarps are aligned along the closely spaced contour lines indicative of nearly vertical slopes (Fig. 3.23). The fault scarps displace the contemporary geomorphic surfaces of late Quaternary–Holocene. Different terraces and fan surfaces have been recognized in the area (Fig. 3.24).

The fault scarp segments represent portions of at least two active faults, the large parts of which are hidden under the fan cover sediments. The sub-window A in Fig. 3.23 is focused on detailed study of active fault features. Figure 3.25 shows a grayscale 3D perspective view of the active fault scarp exposed at locality Talla Syat. The elevation profile P1 constructed across normal to the scarp strikes (Fig. 3.25) indicates ~54 m height consistent with the vertical offset of the faults (Fig. 3.26). The active fault scarps observed in the field are documented in photographs with explanation (Fig. 3.27).

Fig. 3.27 a–d Field
photographs of the active fault
scarps (red dotted line)
showing base of the scarp at
different locations, except
Fp.2, show dip section of the
fault scarp. Locations of Fp.1,
Fp.2, Fp.3, and Fp.4 are
marked in Fig. 3.22. Fp.1:
The red dashed line at the
base of the scarp represents
the active fault trace,
separating the lower surface
(road) and upper surface with
building on the top. Fp.2:
Over the ridge, the red dashed
line defining the NW–SE
trending active fault separates
the lower and upper surfaces
to the left and right sides,
respectively, of the fault trace.
Fp.3: The red dotted line
demarcating the active fault
trace lies between the lower
plain surface in the
foreground and the degraded
fault scarp with farthest trees
lying over the upper surfaces.
Fp.4: The red dashed line
demarcating the base of the
fault scarp with four big
isolated trees over the upper
surface, whereas the man is
standing over the lower
surface

Anaglyph images for Kotadun window (Fault Reference number in the figure is corresponds to characteristics of the fault given in Table 3.6).

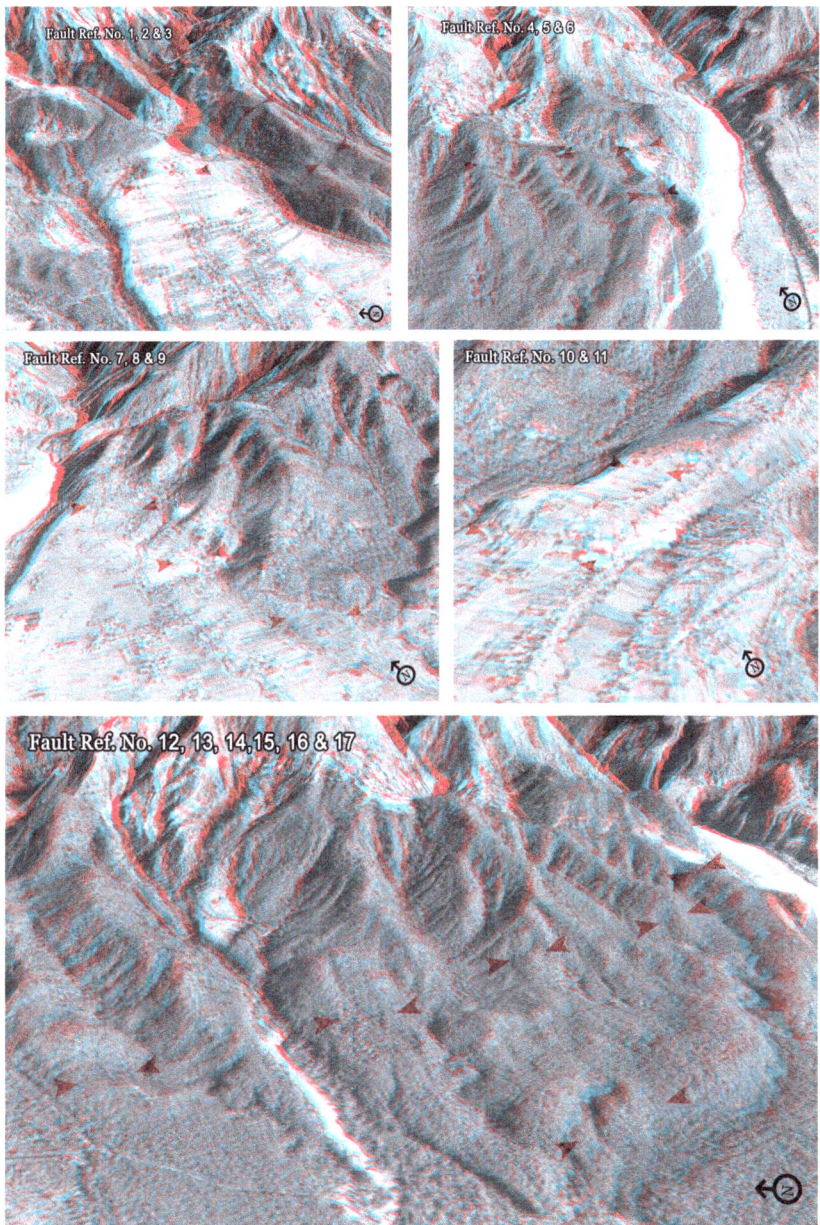

3.4   Active Faults in Ramnagar: Western Kumaun Sub-Himalaya (Kanda Range and Kathgodam)

In western Kumaun Sub-Himalaya zone between the Kanda Range and Kathgodam, three windows-I, II, and III were focused on active fault mapping (Fig. 3.28). The window-I includes the area between the Kanda Range and Ramnagar along the HFT. The window-II area lies along the HFT and the adjoining piedmont zone between Kishanpur and Kaladhungi towns. The window-III encompasses the area near the Kota Dun between the Garjiya Devi and Kota Range.

3.4.1   Ramnagar (Window-I): Kanda Range–Ramnagar

The window-1 covers the area of HFT zone that extends west of the Debka and Kosi rivers across Ramnagar town and further west for several kilometers (Fig. 3.29). Ramnagar is located over the paleofloodplain on the right bank of the Kosi River in the Sub-Himalayan front of western Kumaun. On the western portion, a broad south-verging anticline within the Siwalik Group (Rao et al. 1973) is formed as a fault-propagation fold over the HFT. Fault scarps along the reactivated HFT were mapped on the Sub-Himalayan front. The fault scarps have been generated in the late Quaternary–Holocene sediments byco-seismic offset of the alluvial fan surfaces due to faulting as a result of earthquakes that broke the Himalayan front during the past.

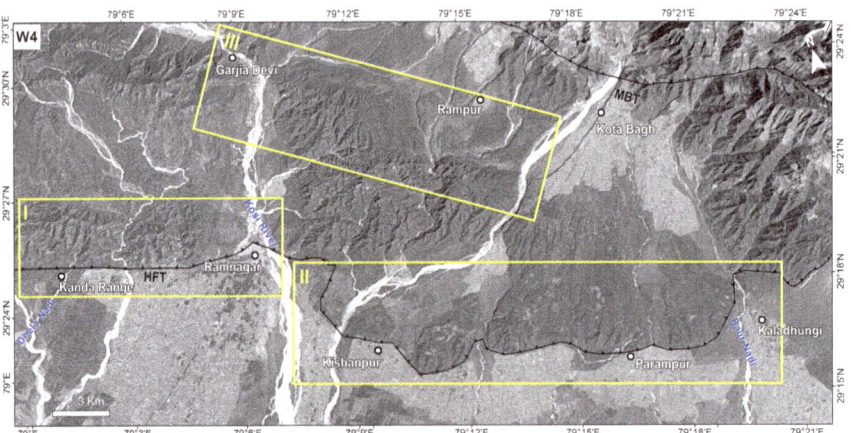

Fig. 3.28 Grayscale map of the Ramnagar region (W 4) shows the physiographic features including fan and terrace surfaces between the MBT and HFT. The roman numerals I, II, and III symbolize selected windows for active fault mapping

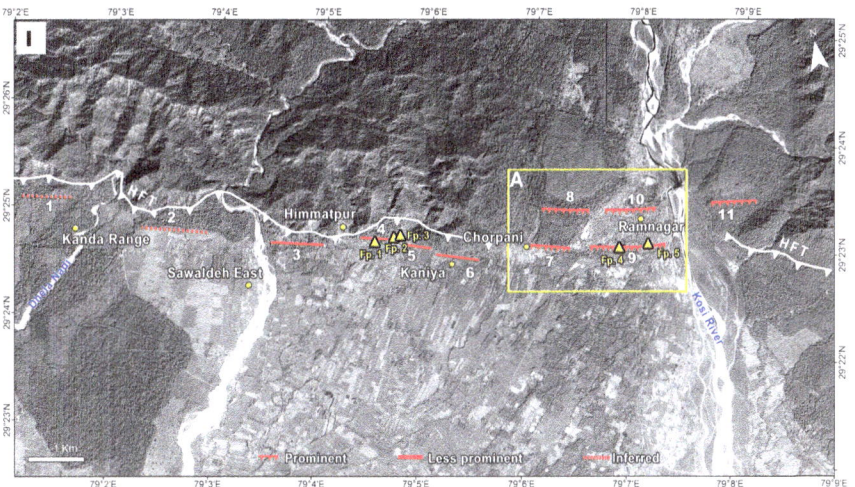

Fig. 3.29 Cartosat-IA gray scale satellite image of the window–I showing the active faults between the Dhela Nadi and Kosi River. Contour and geomorphic map of this region is shown in Figs. 3.30 and 3.31. Numerals on the active fault segments (shown in red color) refer to the fault characteristics given in Table 3.7. Fp.1 to Fp.3 is field photograph locations as shown in Fig. 3.32. The rectangle 'A' is a sub-window with detailed characteristics of the active fault segments shown in Fig. 3.33. Fp.4 to Fp.5 symbolizes field photograph locations as shown in Fig. 3.35

The small segments of the fault identified in the satellite image are assigned numbers 1–11 (Fig. 3.29), and the characteristics of each numbered segment are given in Table 3.7. The topographic control of the scarps is observed in Fig. 3.30 that the fault scarps coincide and align with closely spaced contour lines indicative of slope gradient. The fault scarps are discontinuous and aligned in east–west trend, defining the fault trace of an active fault. Four levels of terraces along the Kosi River and three levels of Kaladhungi fan surfaces were recognized and mapped in the window-I (Fig. 3.31). Field photographs (Fp.1 to Fp.3) of the active fault scarps showing the displacement of alluvial fan surface are shown in Fig. 3.32. In the

Table 3.7 Characteristics of the active fault scarps in Ramnagar (Window-I)

Ref. No.	Thrust	Location	Latitude (N)	Longitude (E)	Strike	Length (m)	Fault feature	Fault reference
1	HFT	Northwest of Kanda Range	29.417040	79.032919	NWW–SEE	830	Fault scarp	Alluvial fan surface
2	HFT	Northwest of Sawal Deh East	29.407734	79.052166	NWW–SEE	1065	Fault scarp	Alluvial fan surface
3	HFT	Southwest of Himmatpur	29.401946	79.071495	NWW–SEE	850	Fault scarp	Alluvial fan surface
4	HFT	East of Himmatpur	29.400233	79.083983	NWW–SEE	500	Fault scarp	Alluvial fan surface
5	HFT	Northwest of Kaniya	29.397873	79.091871	NWW–SEE	334	Fault scarp	Alluvial fan surface
6	HFT	North of Kaniya	29.395190	79.097742	NWW–SEE	598	Fault scarp	Alluvial fan surface
7	HFT	East of Chorpani	29.394076	79.109858	NWW–SEE	624	Fault scarp	River terrace
8	HFT	Northeast of Chorpani	29.398686	79.115153	NWW–SEE	738	Fault scarp	River terrace
9	HFT	South of Ramnagar	29.391372	79.122605	NWW–SEE	1172	Fault scarp	River terrace
10	HFT	North of Ramnagar	29.396970	79.124370	NWW–SEE	792	Fault scarp	River terrace
11	HFT	East of Ramnagar	29.394852	79.142231	NWW–SEE	694	Fault scarp	River terrace

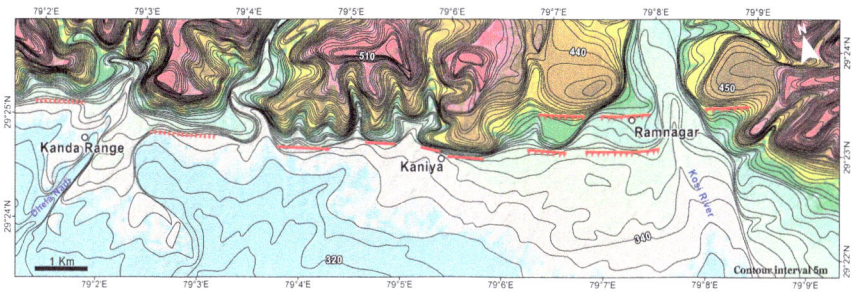

Fig. 3.30 A 5 m interval of contour lines is derived from the Cartosat-IA digital terrain surface showing a fault scarps between Ramnagar and Kanda Range

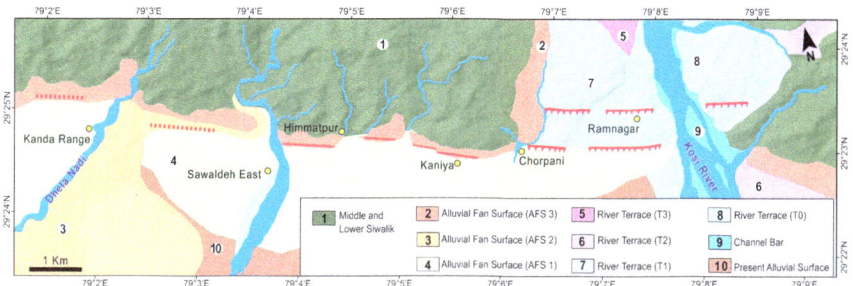

Fig. 3.31 Active fault trace along the HFT in the Ramnagar region (window-I) characterized by active fault scarps that displace the fan surfaces and river terraces between the Dhela Nadi and Kosi River

sub-window 'A' of window-I, we identified fault scarps on the river terraces near the Ramnagar town, through 3D perspective view (Fig. 3.33) and confirmed its existence in finding steep break in surface slopes as indicated in photographs Fp.1 and Fp.2, respectively (Fig. 3.35). A ~12 m of vertical offset was estimated through the construction of elevation profile of the topographic surface across the fault scarp normal to the strike (Fig. 3.34).

Fig. 3.32 **a** Field photograph showing scarp, a person standing is the crest of the scarp, **b** field photograph showing ∼27 m high fault scarp, base of the scarp has been modified by village road construction, **c** field photograph shows fault scarp along the Kaniya shot river cliff section. Field photograph location for Fp.1, Fp.2, and Fp.3 is marked on Fig. 3.29

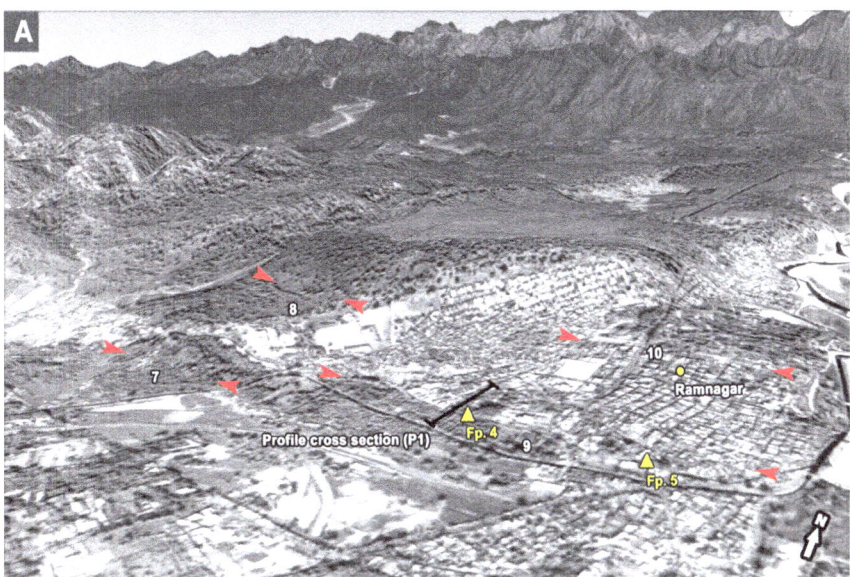

Fig. 3.33 A 3D perspectives view of sub-window A (Fig. 3.29) toward the northern direction showing prominent faults near the Ramnagar town (fault references are 7–10; see Table 3.7). Regional location shown as rectangle 'A' in Fig. 3.29. Thick black line 'P1' showing topographic profile constructed across the active fault is indicated in Fig. 3.34. Digital terrain surface derived from Cartosat-IA stereopair data

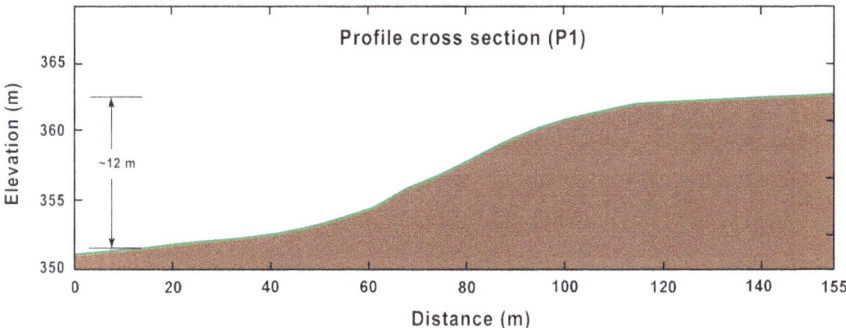

Fig. 3.34 Cross-profile variation of the prominent fault scarp near Ramnagar. Location is marked in Fig. 3.33

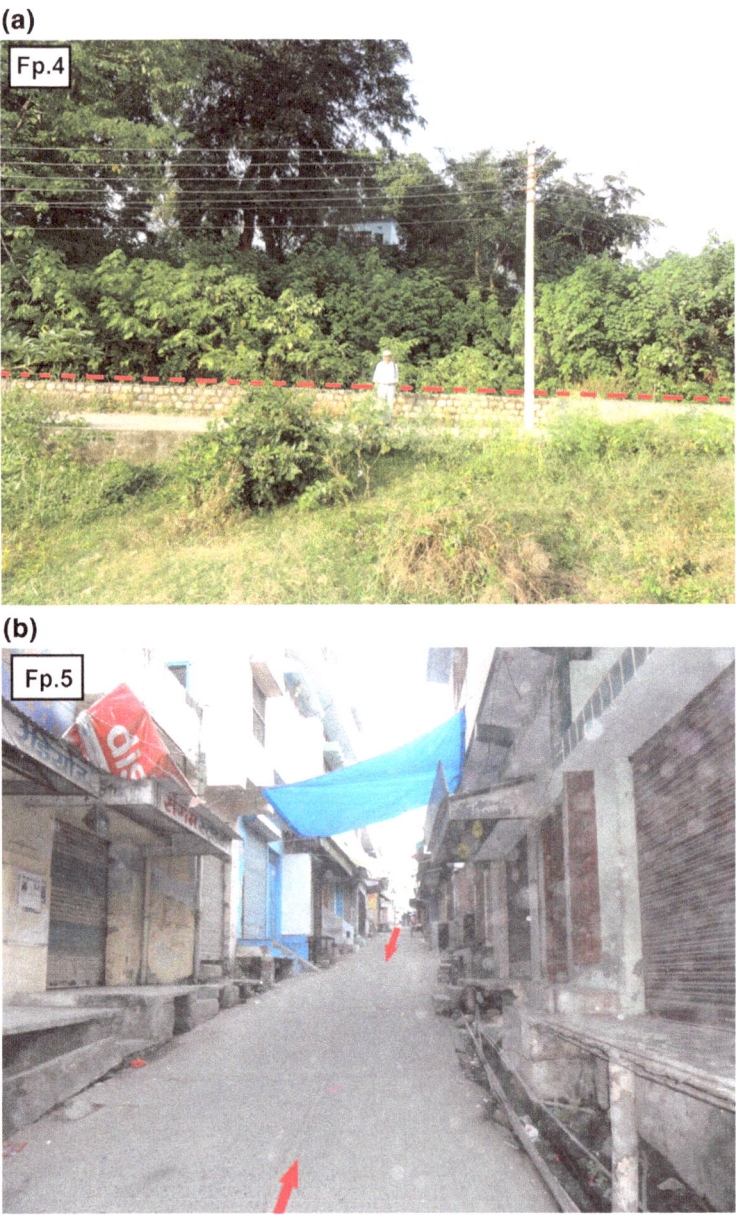

Fig. 3.35 a, **b** Field photographs of the active fault scarps (red dotted line) showing the displacement of river terrace (level T1). Locations of Fp.4 and Fp.5 are marked in Fig. 3.29

Anaglyph images for Ramnagar (window-I). Fault Reference number in the figure is corresponds to characteristics of the fault given in Table 3.7.

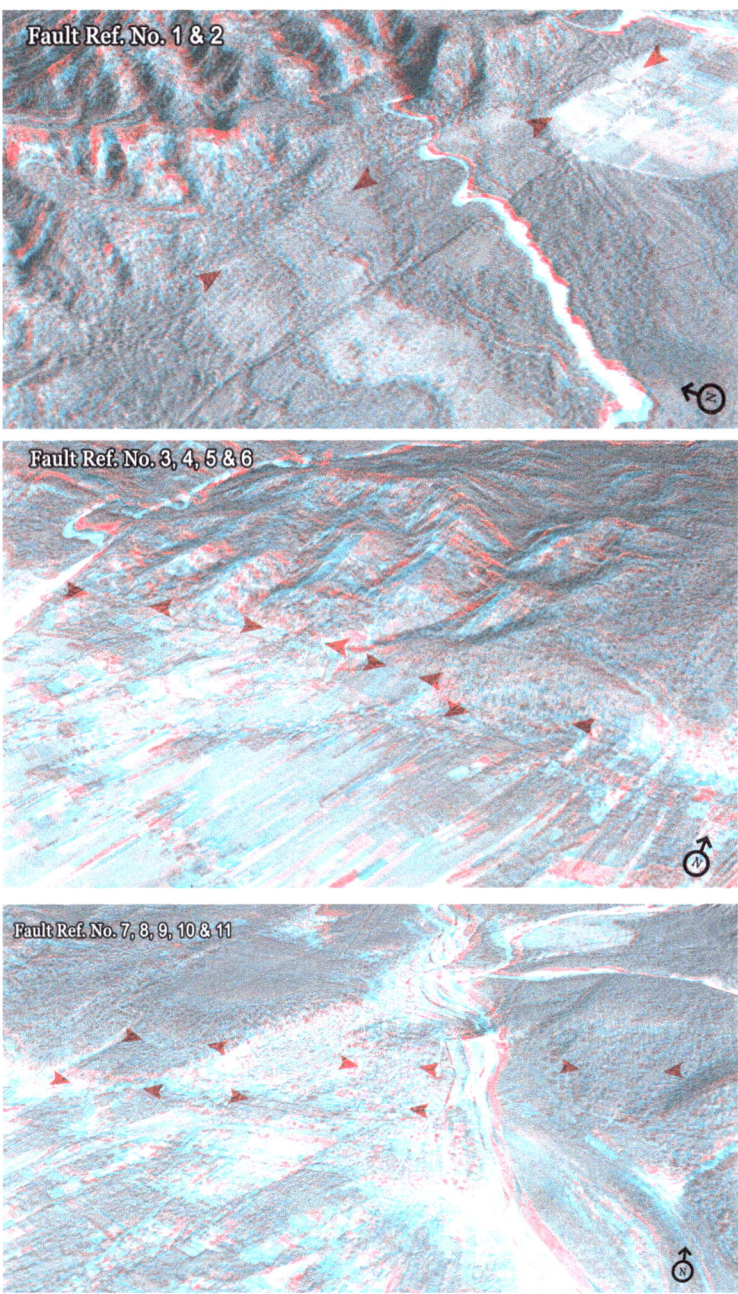

3.4.2 Ramnagar (Window-II): Kishanpur–Kaladhungi

Window-II encompasses the area of HFT zone between near Kaladhungi in the east and Kishanpur in the west. The fault scarps are discontinuous and trend in almost east–west direction (Fig. 3.36). The fault scarps are well expressed in Cartosat-1A imagery. Scarp segments are assigned numbers. Table 3.8 gives characteristics of each fault scarp segment. The topographic expression of the scarps documented in Fig. 3.37 shows that the scarps coincide and are aligned with the steepest slope as indicated in the 5 m interval closely spaced contour lines. Seven levels of river terraces and Kaladhungi piedmont surfaces in the alluvial plain occur in the area. Fault scarp segments affecting different geomorphic surfaces are observed along the Sub-Himalayan front coinciding with the trace of the HFT (Fig. 3.38). This fault scarp is well expressed in three-dimensional perspective view (Fig. 3.39) and elevation profile 'P1' drawn across and normal to the fault scarp which indicates vertical offset (scarp height) of ∼13 m (Fig. 3.40). The trench site is marked as sub-window A of Fig. 3.28, and the trench log details are presented in Fig. 3.41.

Ramnagar Trench

East of Ramnagar, near the village Belparao, a fault scarp ∼13.0 m high of the HFT is exposed. A ∼ 32 m long trench perpendicular to the fault was excavated for paleoseismological investigation (Fig. 3.36). The eastern trench wall was cleaned, and a detailed log on 1 m by 1 m scale was prepared for paleo-earthquake investigation (Kumar et al. 2006) (Fig. 3.41). Five distinct stratigraphic units, 1–5,

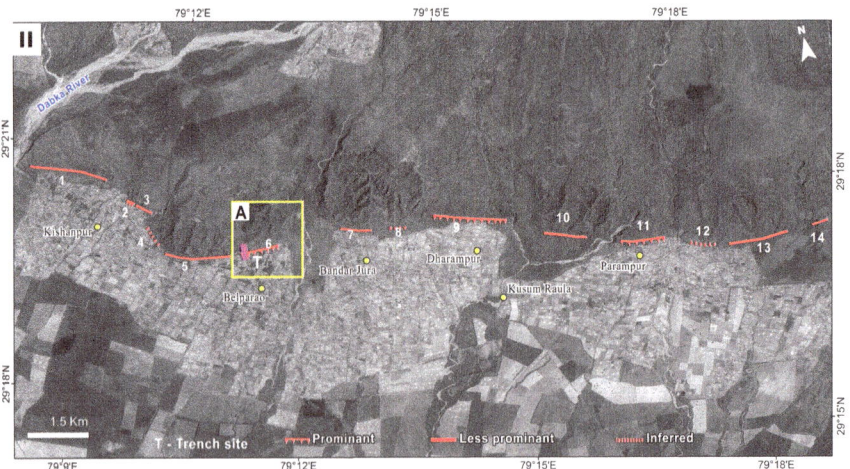

Fig. 3.36 Cartosat-1A grayscale satellite image shows active fault scarps between Kishanpur and near Kaladhungi (Window-II of Ramnagar). Contour and geomorphic map of this region is shown in Figs. 3.37 and 3.38. Numbers on the active faults (shown in red color) refer to characteristics of the fault given in Table 3.8. Rectangle 'A' is a sub-window shows detailed characteristics of the active fault segments (Fig. 3.39). A purple color bar on fault number 6 indicates the trench site

Table 3.8 Characteristics of the Kaladhungi faults

Ref. No	Thrust	Location	Latitude (N)	Longitude (E)	Strike	Length (m)	Fault feature	Fault reference
1	HFT	Northwest of Kishanpur	29.340953	79.167397	NWW–SEE	1676	Kaladhungi fault	Middle and Lower Siwalik Formation
2	HFT	Northeast of Kishanpur	29.331603	79.177452	NW–SE	187	Kaladhungi fault	River terrace
3	HFT	Northeast of Kishanpur	29.329246	79.180285	NW–SE	395	Kaladhungi fault	River terrace
4	HFT	Northeast of Kishanpur	29.323388	79.180123	NNW–SSW	728	Kaladhungi fault	River terrace
5	HFT	Northwest of Belparao	29.317506	79.185734	NWW–SEE	1420	Kaladhungi fault	Middle and Lower Siwalik Formation
6	HFT	North of Belparao	29.316712	79.204057	NEE–SWW	818	Kaladhungi fault	Middle and Lower Siwalik Formation
7	HFT	North of Bandar Jura	29.314421	79.224131	NWW–SEE	693	Kaladhungi fault	Middle and Lower Siwalik Formation
8	HFT	North of Bandar Jura	29.313122	79.233126	NWW–SEE	456	Kaladhungi fault	Middle and Lower Siwalik Formation
9	HFT	North of Dharampur	29.309994	79.248036	NWW–SEE	1562	Kaladhungi fault	River terrace and middle and Lower Siwalik Formation
10	HFT	Northeast of Dharampur	29.302733	79.267696	NWW–SEE	920	Kaladhungi fault	Middle and Lower Siwalik Formation

(continued)

Table 3.8 (continued)

Ref. No	Thrust	Location	Latitude (N)	Longitude (E)	Strike	Length (m)	Fault feature	Fault reference
11	HFT	North of Parampur	29.296496	79.282775	NWW–SEE	948	Kaladhungi fault	River terrace
12	HFT	Northeast of Parampur	29.292589	79.295131	NWW–SEE	540	Kaladhungi fault	River terrace
13	HFT	Northeast of Parampur	29.290682	79.307514	E–W	1241	Kaladhungi fault	Middle and Lower Siwalik Formation
14	HFT	Northeast of Parampur	29.290746	79.321451	E–W	270	Kaladhungi fault	Middle and Lower Siwalik Formation

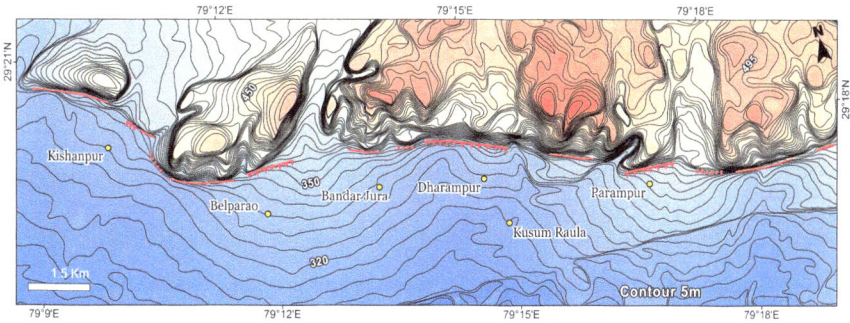

Fig. 3.37 Fault scarps are aligned to the closely spaced contour lines indicating near-vertical slopes. Contour lines (5 m interval) are derived from Cartosat-1A Digital Elevation Model. The geomorphic map of this region is shown in Fig. 3.38

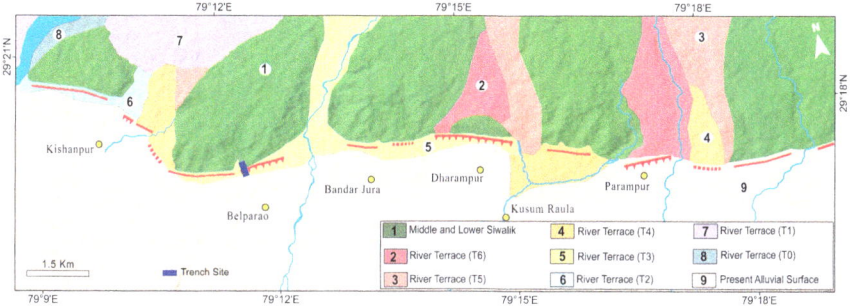

Fig. 3.38 Discontinuous active fault trace of the HFT from Kishanpur to Parampur area (Window-II) affecting the different geomorphic surfaces

were mapped on the trench exposure. The units with conformable contacts are exposed from base to top along the entire length of a trench and dip S at low angle (Fig. 3.41). Unit-1 at the base of the trench consists of rounded to well-rounded, poorly stratified sand and medium boulder gravels. Unit-2 is mottled light tan to yellow sandy clay layer that shows a facies change toward the northern end of the trench. Unit-3 overlying unit-2 is clayey medium to coarse-grained sand with occasional discontinuous gravel stringers. Unit-4 is clayey silt to coarse sand with well-defined thin channels. It is oxidized toward top and shows red colorization.

The uppermost unit-5 consists of dark clayey sand package that shows weak soil structure with few randomly oriented pebbles and cobbles. The youngest unit-5 is colluvium derived from the scarp. Three fault strands, F1, F2, and F3, were mapped on the trench exposure that shows displacement of units 1–4. A detrital charcoal sample BR-09 from faulted unit-2 provides an age of A.D. 984–1158 (Fig. 3.41). Samples from faulted units 3 and 4 give radiometric ages of A.D. 1259–1390 and A.D. 1278–1390, respectively (samples BR-15 and BR-06; Fig. 3.41, top right). The unfaulted unit-5 of colluvium postdates faulting. Detrital charcoal (sample

Fig. 3.39 A three-dimensional view of sub-window A (Fig. 3.36) toward northern direction showing prominent faults in the north of Belparao (red arrow). Number denotes fault reference as shown in Table 3.8. Thick black line P1 showing topographic profile constructed across the active fault is indicated in Fig. 3.40. A 5 m digital terrain surface derived from Cartosat-1A stereopair data

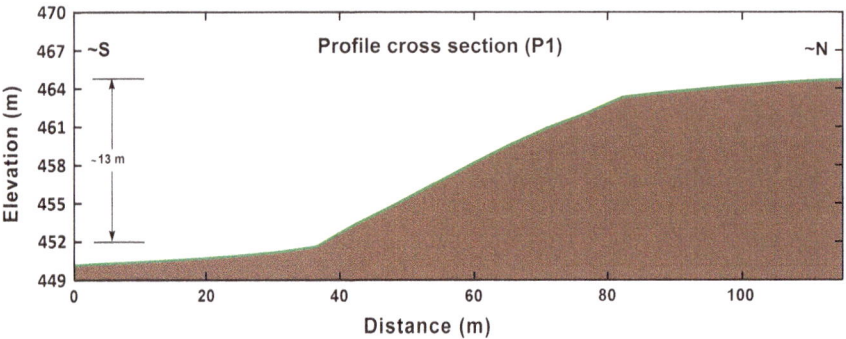

Fig. 3.40 Topographic profile across a prominent fault scarp. Location marked in Fig. 3.39

BR-07; Fig. 3.41 top right) from this colluviums provides an age of A.D. 1301–1433 that limit the most recent displacement. Sample BR-07 from unit-5 may be reworked, and hence upper bound age may not be firm and imply older than the deposit in which it sits.

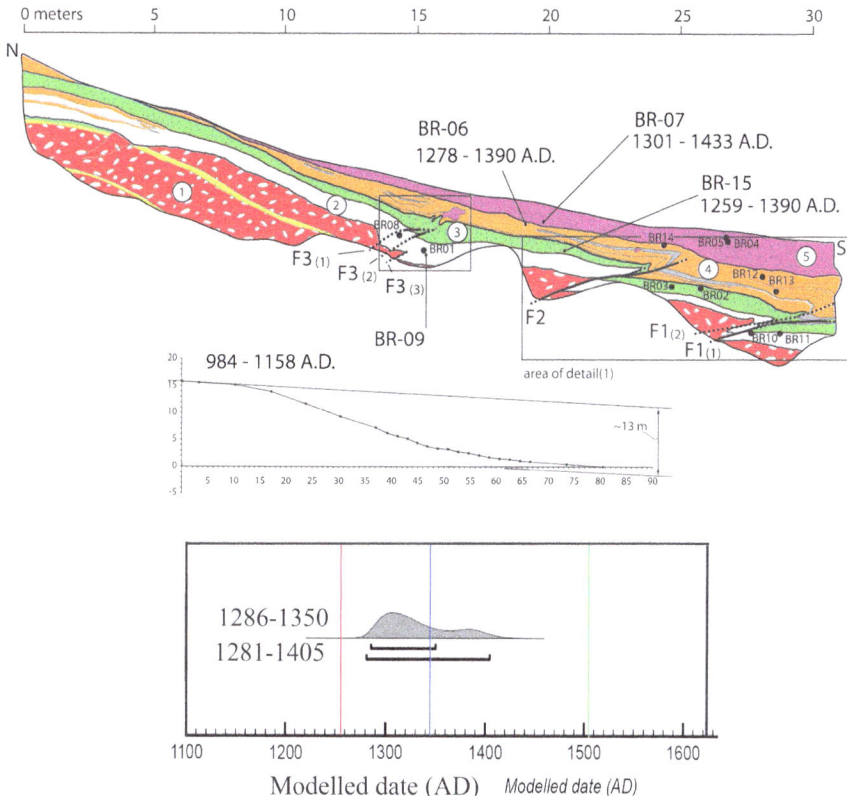

Fig. 3.41 Top: Ramnagar trench logs and fault scarp elevation profile. Bottom: One- and two-sigma confidence intervals are given for rupture dates at Ramnagar site, alongside probability distribution functions developed from radiocarbon age results using the OxCal software program (Bronk Ramsey 2009a). Vertical lines indicating dates of putative earthquakes in 1255 CE (red vertical line), 1344 CE (blue vertical line), and 1505 CE (green vertical line) are superimposed. Radiocarbon sample collection locations are shown in a trench log provided in top panel (Jayangondaperumal et al. 2017)

Based on trench data, it was inferred that the total brittle displacement is ∼4 m recorded along fault traces F1, F2, and F3 due to a single earthquake event. The timing of earthquake is constrained between post-A.D. 1278 and pre-A.D. 1278. More recently, modeling of published age data of Ramnagar trench (Kumar et al. 2006) using Bayesian statistical program OxCal indicates an event, coeval on the HFT, which corresponds to a historically documented A.D. 1344 earthquake reported from Nepal (Pant 2002) with an estimated magnitude of >8.6 Mw (Jayangondaperumal et al. 2017).

Anaglyph images for Kishanpur–Kaladhungi (Ramnagar: Window-II). Fault Reference number in the figure is corresponds to characteristics of the fault given in Table 3.8.

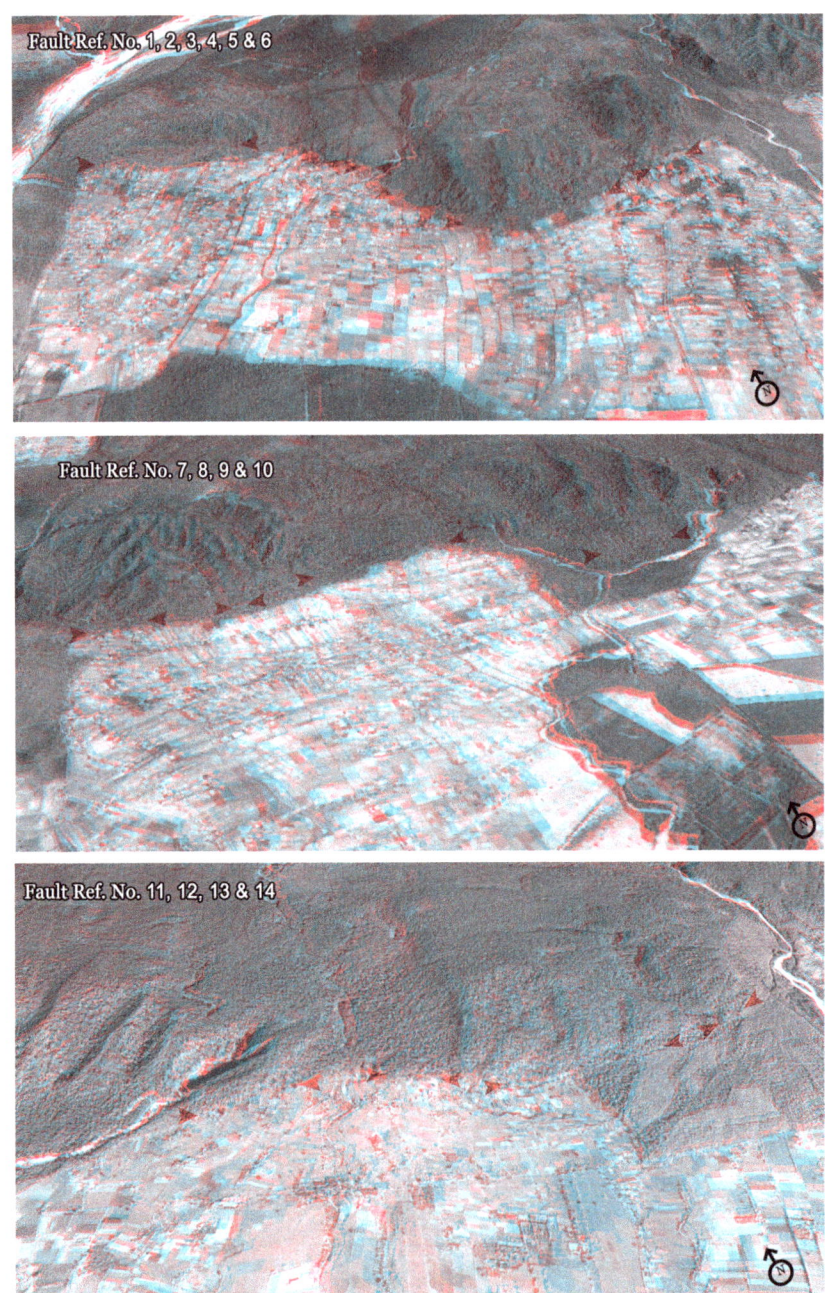

3.4.3 Ramnagar (Window-III): Garjiya Devi and Kota Range

The Window-III of Ramnagar includes the area the southern portion of the Kota Dun, north of the HFT zone between Garjiya Devi and Kota Range. Four active segments of fault scarps trending ∼E–W to NE–SW are exposed at around Garjiya Devi and south of Bandarpani. We define a fault connecting the fault segments and call it as Bandarpani Fault, and spring lies along this fault. The fifth fault segment is observed to the north of the Kota Range. The numerals 1–5 on the fault segments shown on Cartosat-1A imagery (Fig. 3.42) refer to the fault characteristics given in Table 3.9. The topographic scarp represents the fault scarp characterized by steep slopes shown by closely spaced 10 m interval contour lines (Fig. 3.43). The fault contact lies between the Siwalik strata and alluvial gravels (Fig. 3.44). A 3D perspective image is constructed for active fault scarp northeast of Bhangajala Nadi shows a north-facing fault scarp characterized by a down–faulted northern block in a narrow half-graben suggesting normal faulting (Fig. 3.45). The location of elevation profile 'P1' is shown in Fig. 3.42. Profile constructed across and normal to the scarp numbered 3 indicates vertical separation of ∼20 m (Fig. 3.46).

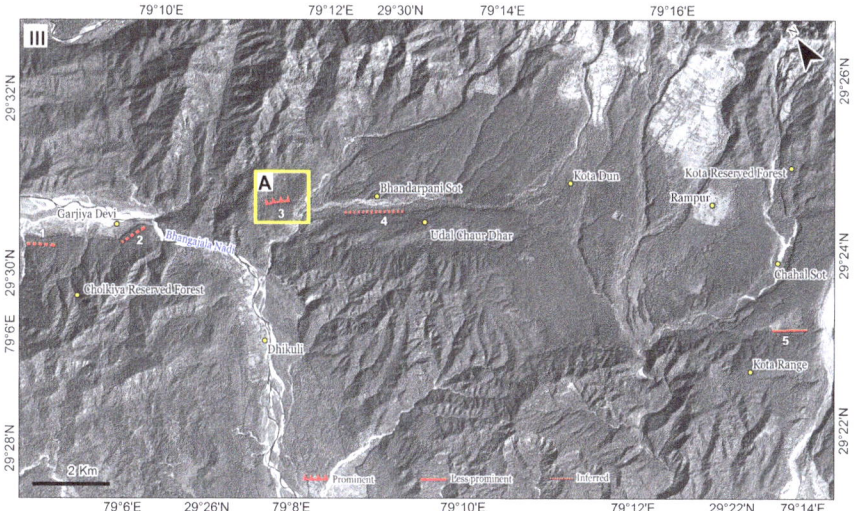

Fig. 3.42 Cartosat-IA grayscale satellite image shows active faults between Garjiya Devi and Kota Range (Window-III of Ramnagar). Contour and geomorphic map of this region is shown in Figs. 3.43 and 3.44. Numerals on the active fault (shown in red color) refer to fault characteristics given in Table 3.9. Rectangle A is a sub-window location where detailed characteristics of active faults studied (Figs. 3.45 and 3.46)

Table 3.9 Characteristics of the back tilting fault scarps between Garjiya Devi and Kota Range

Ref. No	Thrust	Location	Latitude (N)	Longitude (E)	Strike	Length (m)	Fault feature	Fault reference
1	HFT	Southwest of Garjiya Devi	29° 30′ 19.18″N	79° 06′ 58.94″E	NW–SE	659	North-facing fault scarp	Kotabagh surface
2	HFT	South of Garjiya Devi	29° 29′ 42.66″N	79° 08′ 04.47″E	NWW–SEE	634	North-facing fault scarp	Kotabagh surface
3	HFT	Northeast of Bhangajala Nadi	29° 28′ 50.41″N	79° 10′ 2.89″E	NW–SE	582	North-facing fault scarp	Kotabagh surface
4	HFT	South of Bhandarpani Sot	29° 28′ 00.51″N	79° 11′ 05.42″E	NW–SE	1398	North-facing fault scarp	Hill slope
5	HFT	Northeast of Kota Range	29° 23′ 29.61″N	79° 15′ 05.06″E	NW–SE	850	North-facing fault scarp	Hill slope

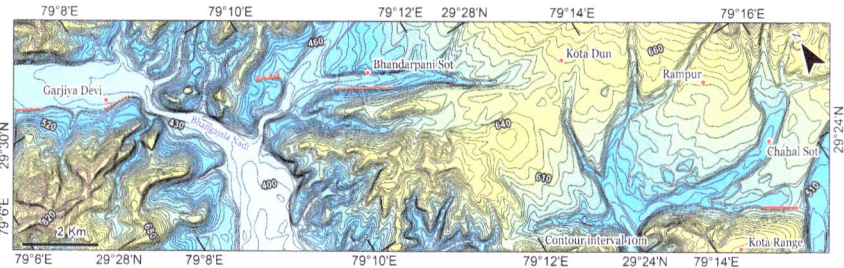

Fig. 3.43 Active fault scarps are aligned along the closely spaced contour lines indicate steepness of near-vertical slopes. Contour interval of 10 m is derived from Cartosat-1A Digital Elevation Model. The geomorphic map of this region is shown in Fig. 3.44

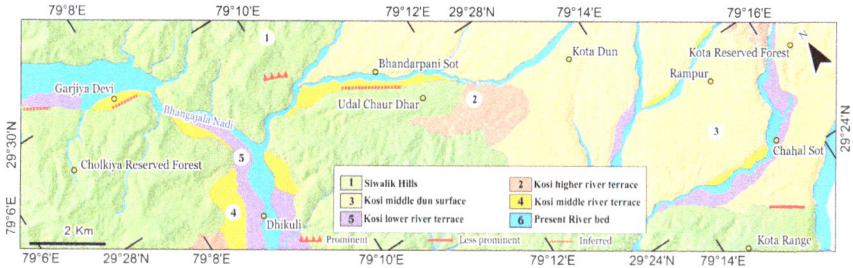

Fig. 3.44 Active fault traces are demarcated along the contact between the Siwaliks (green), the fan surface, and river terraces, showing north-facing fault scarps

Fig. 3.45 A three-dimensional perspectives view of rectangle A (Fig. 3.42) toward the northern direction showing a north-facing fault scarp (red arrow) to the east of Garjiya Devi. Numerals are fault reference shown in Table 3.9. Thick black line 'P1' denotes the profile cross section constructed across the active fault and is given in Fig. 3.46. Digital Elevation Model is derived from Cartosat-1A stereopair data

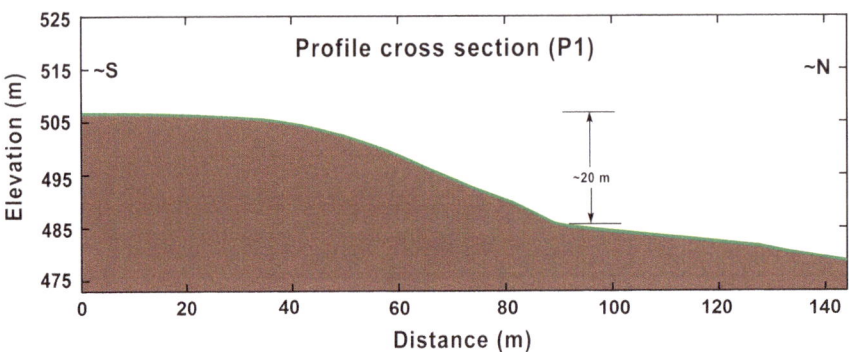

Fig. 3.46 Topographic profile across the prominent fault scarp of 3. Location is marked in Fig. 3.45

Anaglyph images for active faults along the Garjiya Devi and Kota Range (Ramnagar: Window-III). Fault Reference number in the figure is corresponds to characteristics of the fault given in Table 3.9.

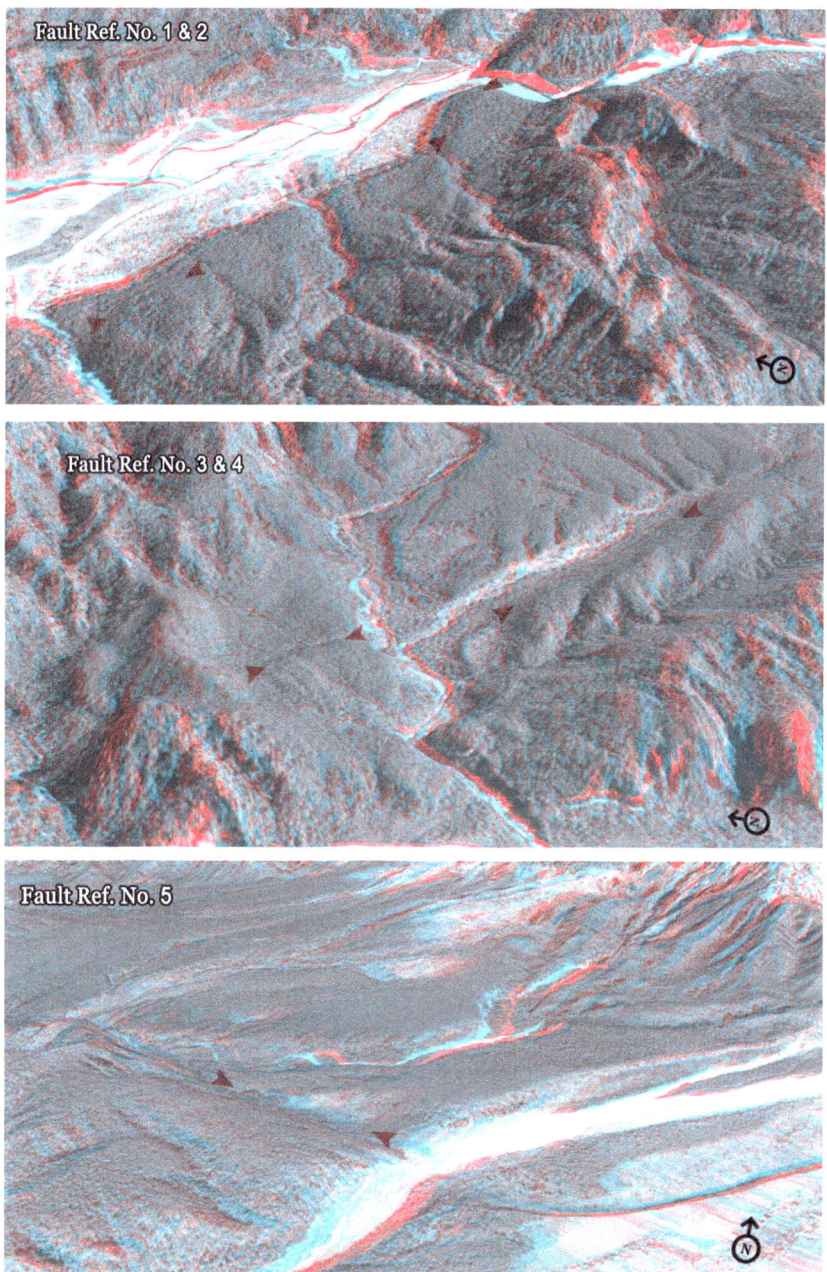

3.5   Active Faults Between Bhimtal and Khatima

In the area, the physiographic break between the Sub-Himalayan front and the Gangetic Plain, and the tectonic contact between the Siwalik Group strata and the alluvium demarcate the trace of the HFT. The HFT was developed regionally during Quaternary period (Valdiya 1992; Mugnier et al. 2005; Thakur et al. 2007). We have mapped fault scarps in contemporary surfaces made of Holocene–Recent or late Quaternary sediments along the HFT trace. The fault scarps represent active faulting implying reactivation of the HFT. We have focused active fault mapping on three windows I, II, III under Window 6 (Fig. 3.47).

3.5.1   Window-I: Logar Gad

At Logar Gad in the Gaula Valley, a mountain front fan has been faulted by an active fault along the MBT (Valdiya 1992; Kothyari et al. 2010; Philip et al. 2017). The active faulting has resulted in ground rupture forming a NE-facing fault scarp measuring about 1.8 km in length and 21 m in height (Fig. 3.48 and Table 3.9). Numerals in Fig. 3.48 refer to fault characteristics given in Table 3.10. The

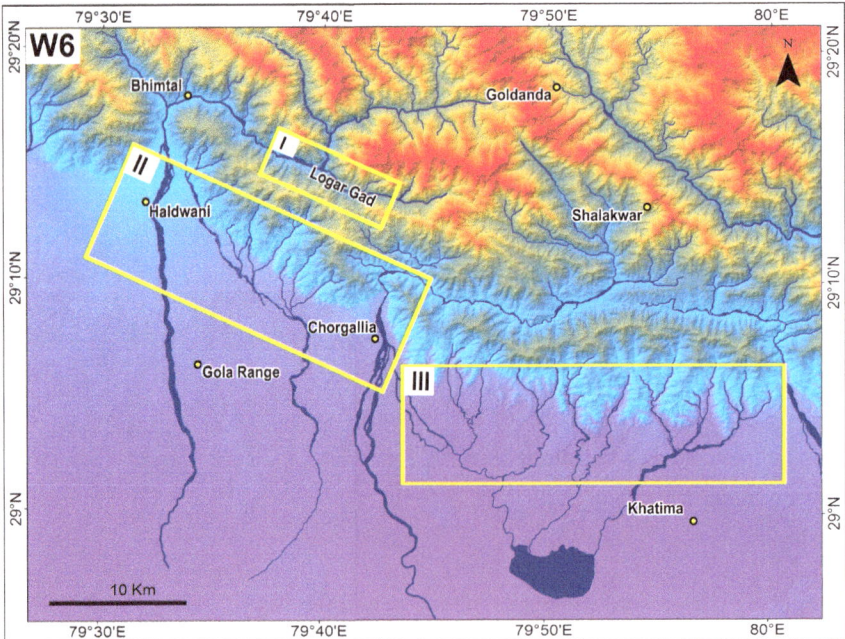

Fig. 3.47 Digital elevation surface shows selected windows for mapping active faults between Bhimtal and Khatima (Window 6) in Kumaun Sub-Himalaya

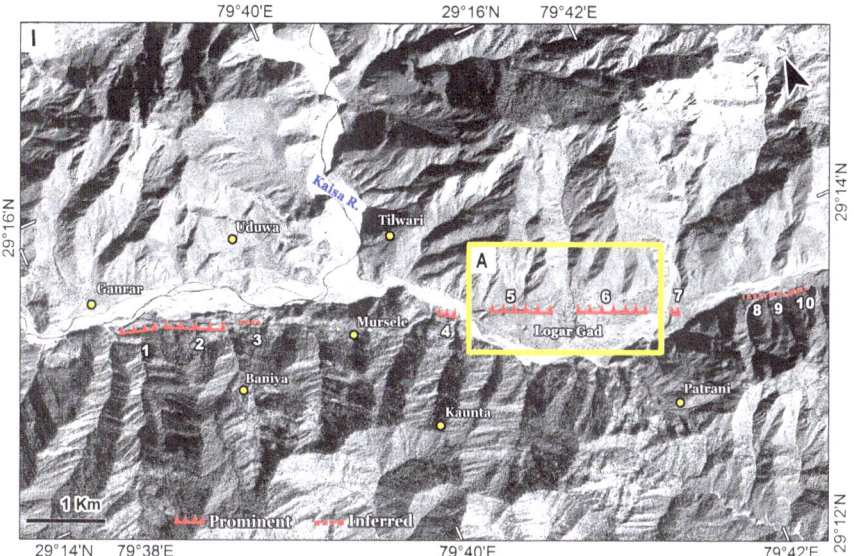

Fig. 3.48 Cartosat-1A grayscale satellite imagery of Window-I showing the active faults between Ganrar and Logar Gad. Contour and geomorphic map of this region is shown in Figs. 3.49 and 3.50. Numerals on active faults (red color) refer to fault characteristics given in Table 3.10. Rectangle A is a sub-window location (Fig. 3.51) where detailed characteristics of active faults studied (Figs. 3.51 and 3.52)

sub-window A showed in Fig. 3.48 shows close-up view of the active fault at Logar Gad. The mountain front north of the active fault is made of quartzite—volcanics of the Lesser Himalaya. The active fault in the fan sediments is an east–west continuation of the MBT. Logar Gad is a type example that shows reactivation of the MBT with normal fault sense of motion. The footwall of the MBT comprises steeply dipping sandstone of the Logar Formation (i.e., Dharamsala Formation, Raiverman 2002), which in turn overthrusts the Siwalik Group along the NE-dipping Logar Thrust (Raiverman 2002; Luirei et al. 2014; Philip et al. 2017), which in our interpretation is an equivalent of the Medlicott-Wadia Thrust (Thakur et al. 2010). The fan deposits consist of quartzite and volcanic clasts derived from the Lesser Himalaya north of the MBT. Further, west of Logar Gad, between Mursele and Ganrar, an active fault trace (1, 2, and 3) is demarcated on the fan surface defined by a north-facing scarp representing the western continuation of the active fault at the Logar Gad. The active faults mapped in this study coincide with the trace of MBT, indicating its reactivation. Active faults between Ganrar and Patrani are aligned along closely spaced contour intervals suggesting a scarp

Table 3.10 Characteristics of the Logar Gad fault scarps

Ref. No	Thrust	Location	Latitude (N)	Longitude (E)	Strike	Length (m)	Fault feature	Fault reference
1	Logar Gad Thrust	Southeast of Ganrar	29° 15' 03.45"N	79° 38' 31.25"E	NNE–SSW	430	Fault scarp	Terrace
2	Logar Gad Thrust	Southwest of Uduwa	29° 14' 55.21"N	79° 38' 52.43"E	NE–SW	702	Fault scarp	Terrace
3	Logar Gad Thrust	North of Baniya	29° 14' 48.30"N	79° 39' 14.02"E	NE–SW	212	Fault scarp	Terrace
4	Logar Gad Thrust	Northeast of Mursele	29° 14' 18.16"N	79° 40' 29.54"E	NE–SW	268	Fault scarp	Terrace
5	Logar Gad Thrust	Northeast of Kaunta	29° 14' 08.51"N	79° 40' 53.60"E	NE–SW	940	Fault scarp	Terrace
6	Logar Gad Thrust	Northwest of Patrani	29° 13' 52.08"N	79° 41' 31.50"E	NE–SW	920	Fault scarp	Terrace
7	Logar Gad Thrust	North of Patrani	29° 13' 41.46"N	79° 41' 54.27"E	NNE–SSW	97	Fault scarp	Terrace
8	Logar Gad Thrust	Northeast of Patrani	29° 13' 34.98"N	79° 42' 25.59"E	NNE–SSW	261	Fault scarp	Terrace
9	Logar Gad Thrust	Northeast of Patrani	29° 13' 31.65"N	79° 42' 34.86"E	NE–SW	136	Fault scarp	Terrace
10	Logar Gad Thrust	Northeast of Patrani	29° 13' 29.69"N	79° 42' 43.97"E	N–S	277	Fault scarp	Terrace

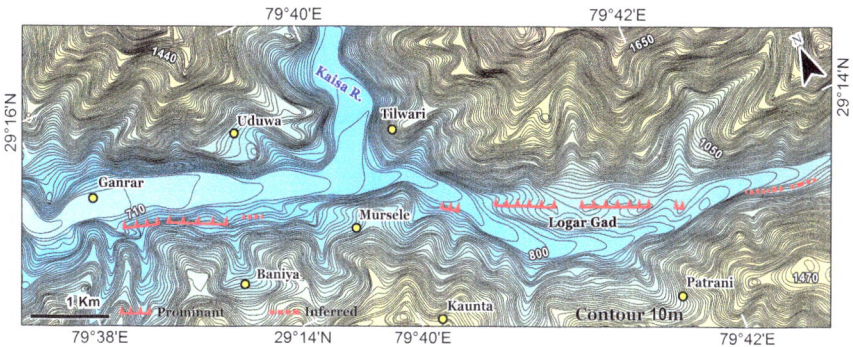

Fig. 3.49 Active fault scarps are aligned along the closely spaced contour lines indicate near-vertical slopes. Contour lines (10 m interval) are derived from Cartosat-I digital terrain surface

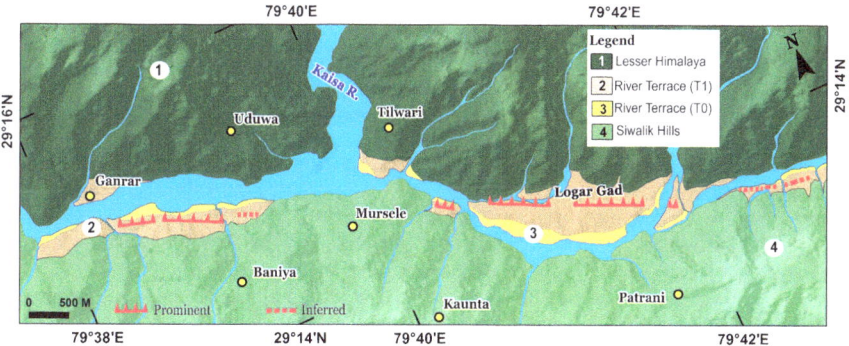

Fig. 3.50 North-facing active fault traces in the Ganrar–Logar Gad area (Window-I) traversing the river terraces

(Fig. 3.49). Here, the active fault along Logar Gad follows the curvature of the contour lines. A three-dimensional perspective view was created for the Logar Gad fan, in which a north-facing scarp affecting the fan surface is well expressed (Fig. 3.51). A topographic profile P1, location shown in Fig. 3.51, constructed normal to Logar Gad fault scarp indicates a vertical offset of ~12 m (Fig. 3.52).

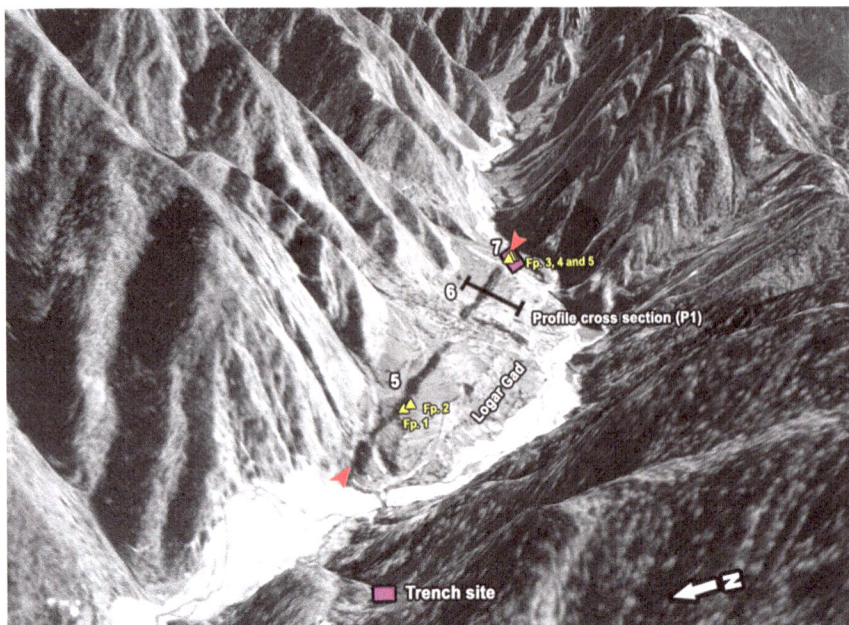

Fig. 3.51 Three-dimensional perspective view with Cartosat-1A overlays showing north–facing fault in the Logar Gad. Red arrows mark the active fault dissecting the river terrace. Numerals denoting fault reference are shown in Table 3.10 and Fig. 3.48. Regional location shown as rectangle 'A' in Fig. 3.48. Thick black line 'P1' showing topographic profile constructed across the active fault is indicated in Fig. 3.52. Yellow triangles indicate field photographs as shown in Fig. 3.53. Field digital terrain surface derived from Cartosat-1A stereopair data

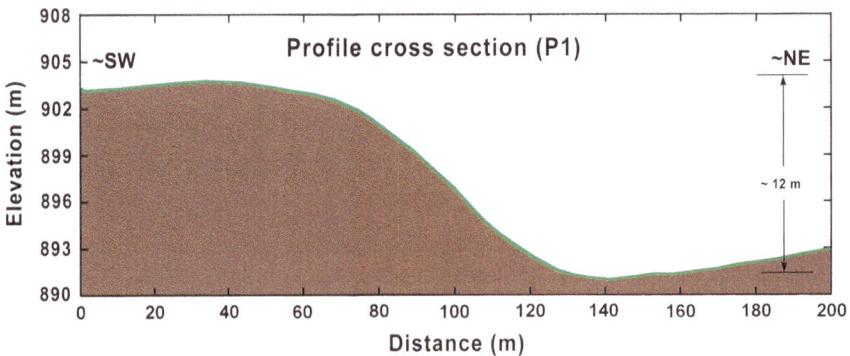

Fig. 3.52 Elevation profile across the fault scarp in fan surface indicates ∼ 12 m vertical offset near Logar Gad. Location is marked in Fig. 3.51

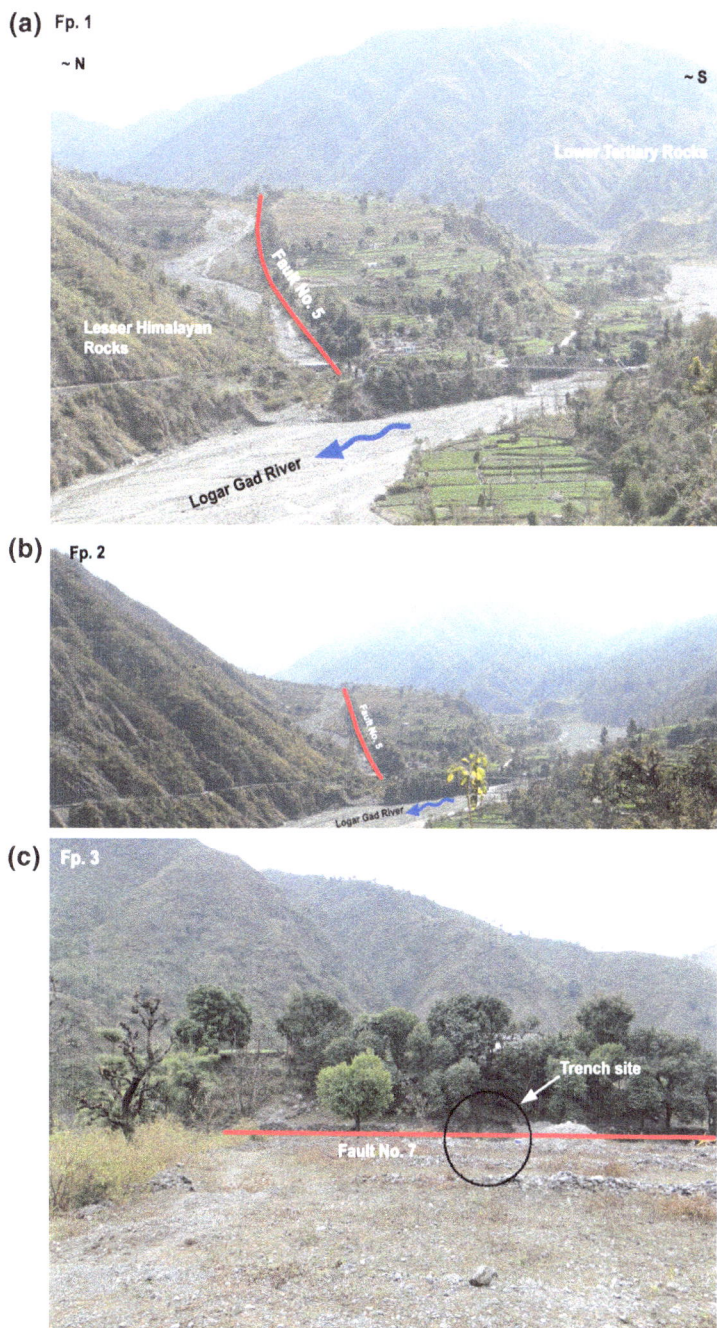

Fig. 3.53 Field photographs of the active fault scarps (fault no. 5 and 7, see Table 3.10) showing displacement of the fan surfaces at different locations. Locations are marked in Fig. 3.51. Field photographs Fp.1 and Fp.2 indicate fault no. 5, and field photographs Fp.3, Fp.4, and Fp.5 indicate fault no. 7

Fig. 3.53 (continued)

Anaglyph images for active faults in Logar Gad (Window-I). Fault Reference number in the figure is corresponds to characteristics of the fault given in Table 3.10.

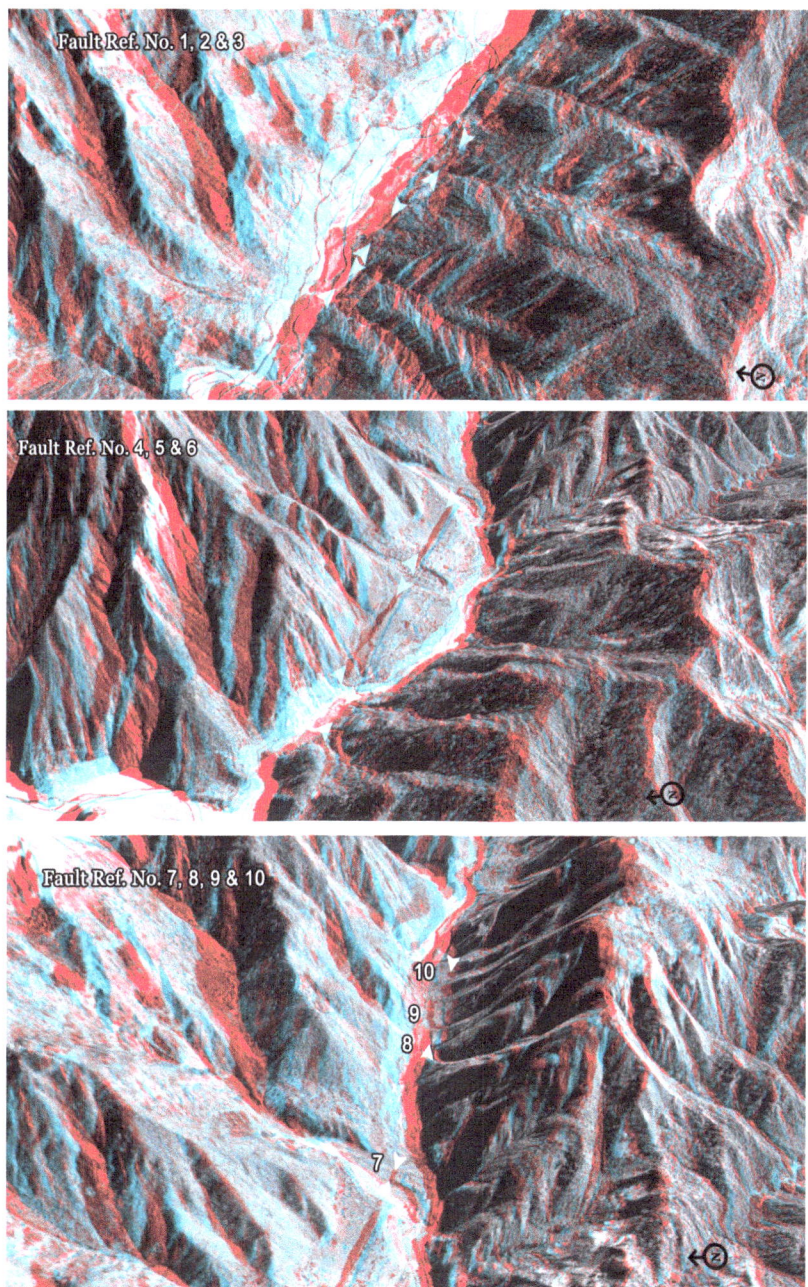

3.5.2 Window-II: Haldwani to Chorgallia

Between the towns of Haldwani and Chorgallia, the HFT trending WNW–ESE is defined by an abrupt physiographic break between the Sub-Himalayan front and the alluvial plain. The fault scarps are well expressed in the satellite imagery as topographic break in slope (Fig. 3.54). Locality-wise fault (1–6) characteristics are given in Table 3.11 (Fig. 3.54). Scarps defining the active faults are aligned along the closely spaced contour intervals implying steepness of the slopes (Fig. 3.55). Active faults are manifested in the form of fault scarps which were formed by displacement of the contemporary fan surfaces (Fig. 3.56). North of Chorgallia, the fault scarp displaces the piedmont plain as well as the river terraces. At other places west of Chorgallia, the fault scarps occur on the Himalayan front contact with the piedmont and alluvial plain. A three-dimensional perspective map prepared from DEM and Cartosat-1A imagery of the Chorgallia area displays a well-marked fault scarp (Fig. 3.57). Northwest of Chorgallia, a topographic profile was constructed normal to the fault scarp (location shown in Fig. 3.57). The profile depicts a scarp of ∼36 m height, suggesting near value of the fault offset (Figs. 3.58 and 3.59).

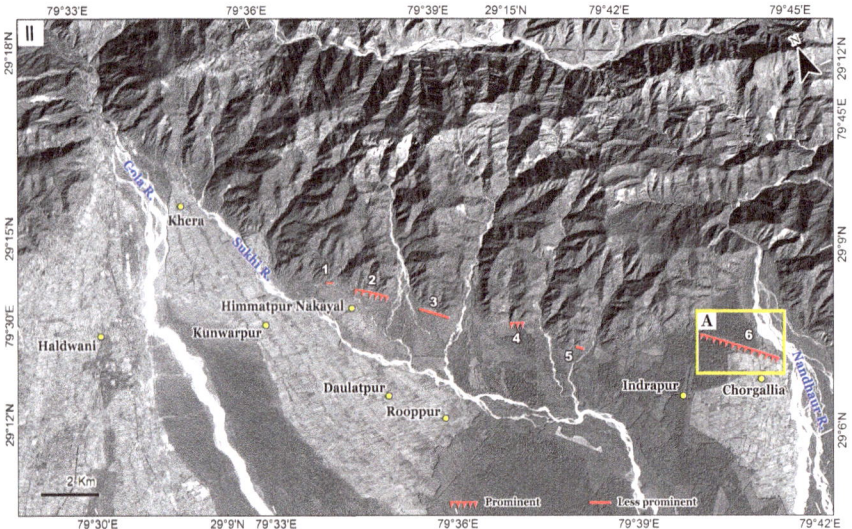

Fig. 3.54 Cartosat-I grayscale image of Window-II shows active fault traces between Haldwani and Chorgallia. Contour and geomorphic map of this region is shown in Figs. 3.55 and 3.56. Numerals on active fault (shown in red color) refer to fault characteristics of the fault given in Table 3.11. Rectangle A is a sub-window (Fig. 3.57) location where detailed characteristics of active faults studied

Table 3.11 Characteristics of the fault scarps between Haldwani and Chorgallia

Ref. No	Thrust	Location	Latitude (N)	Longitude (E)	Strike direction	Length (m)	Fault feature	Fault reference
1	HFT	Northwest of Himmatpur Nakayal	29° 11' 59.85"N	79° 35' 30.84"E	NW–SE	223	Fault scarp	Surface
2	HFT	Northeast of Himmatpur Nakayal	29° 11' 31.23"N	79° 36' 07.84"E	NW–SE	1036	Fault scarp	Hill slope/terrace
3	HFT	Northeast of Daulatpur	29° 10' 47.85"N	79° 37' 04.61"E	NW–SE	1000	Fault scarp	Surface
4	HFT	Northeast of Rooppur	29° 10' 00.58"N	79° 38' 19.65"E	NNW–SSE	408	Fault scarp	Hill slope/surface
5	HFT	Northwest of Indrapur	29° 09' 09.72"N	79° 39' 11.20"E	NW–SE	223	Fault scarp	Surface
6	HFT	North of Chorgallia	29° 07' 59.29"N	79° 41' 52.17"E	NW–SE	2584	Fault scarp	Surface

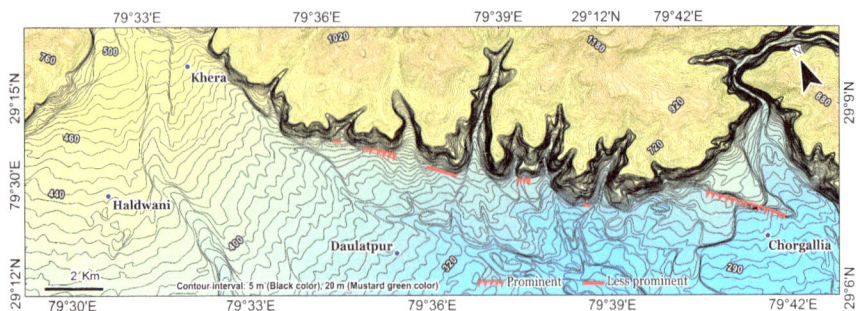

Fig. 3.55 Active fault scarps are aligned along the closely spaced contour lines indicating near-vertical slopes. Contour lines in fan surface show 5 m interval (black tone), and contour lines in Siwalik Formation show 20 m interval (mustard green tone). Contour lines are derived from Cartosat-1A DEM of 10 m resolution

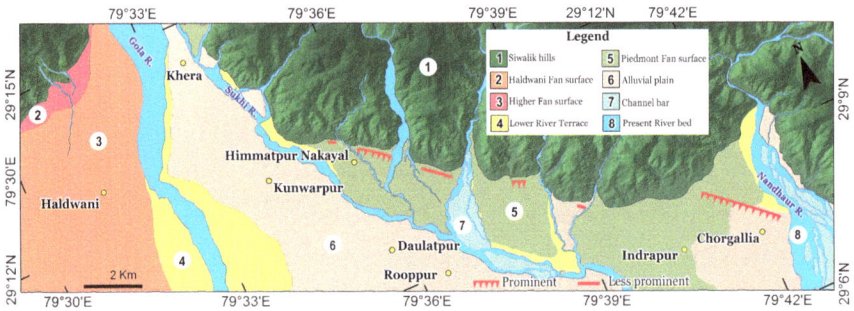

Fig. 3.56 Active fault traces of the HFT in the Window-II between Haldwani and Chorgallia

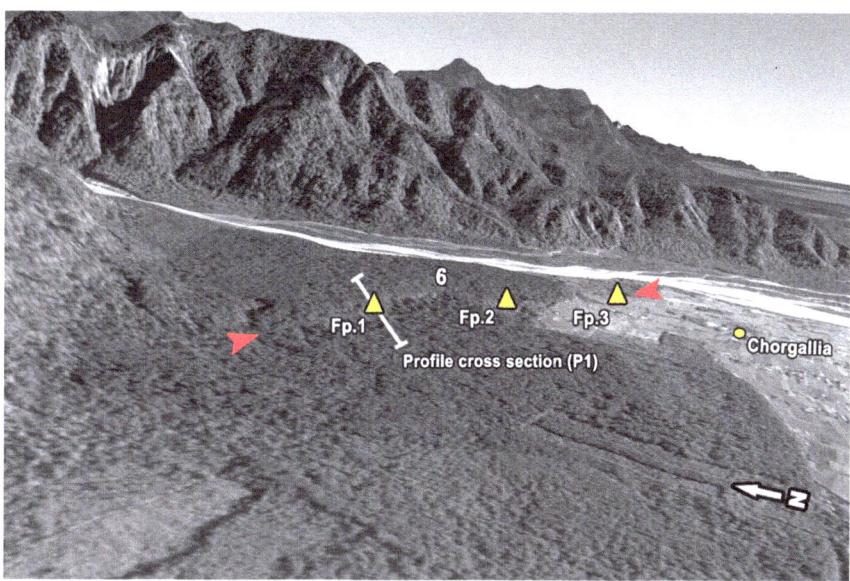

Fig. 3.57 A three-dimensional perspective view of the active fault trace (red arrows) in Chorgallia shown in Cartosat-1A overlaid DEM image (fault reference 6, see Table 3.11). Regional location shown as rectangle A in Fig. 3.54. Thick black line P1 showing topographic profile constructed across the active fault is indicated in Fig. 3.58. A 5 m digital terrain surface derived from Cartosat-1A stereopair data

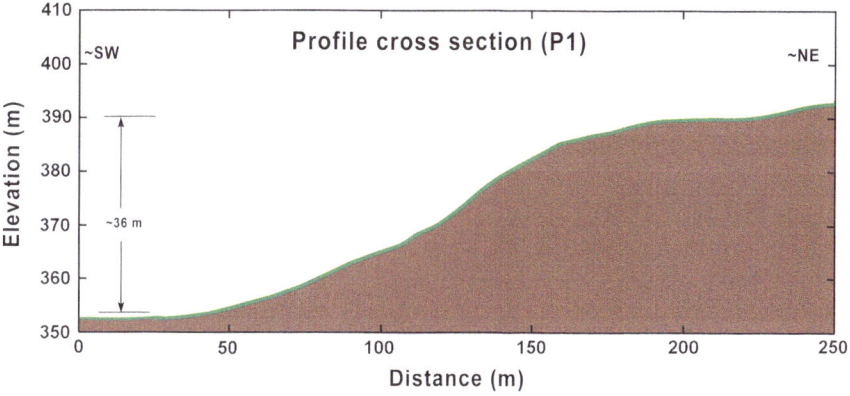

Fig. 3.58 Profile P1 across the fault scarp in piedmont fan surface near Chorgallia indicates 36 m vertical offset. Profile location is marked in Fig. 3.57

Fig. 3.59 Field photograph showing fault scarp at location Fp.1, Fp.2, and Fp.3. Location of field photograph is shown in 3.57. Discontinuous line shows the base of the fault scarp

Anaglyph images for active faults between Haldwani and Chorgalia (Window-II). Fault Reference number in the figure is corresponds to characteristics of the fault given in Table 3.11.

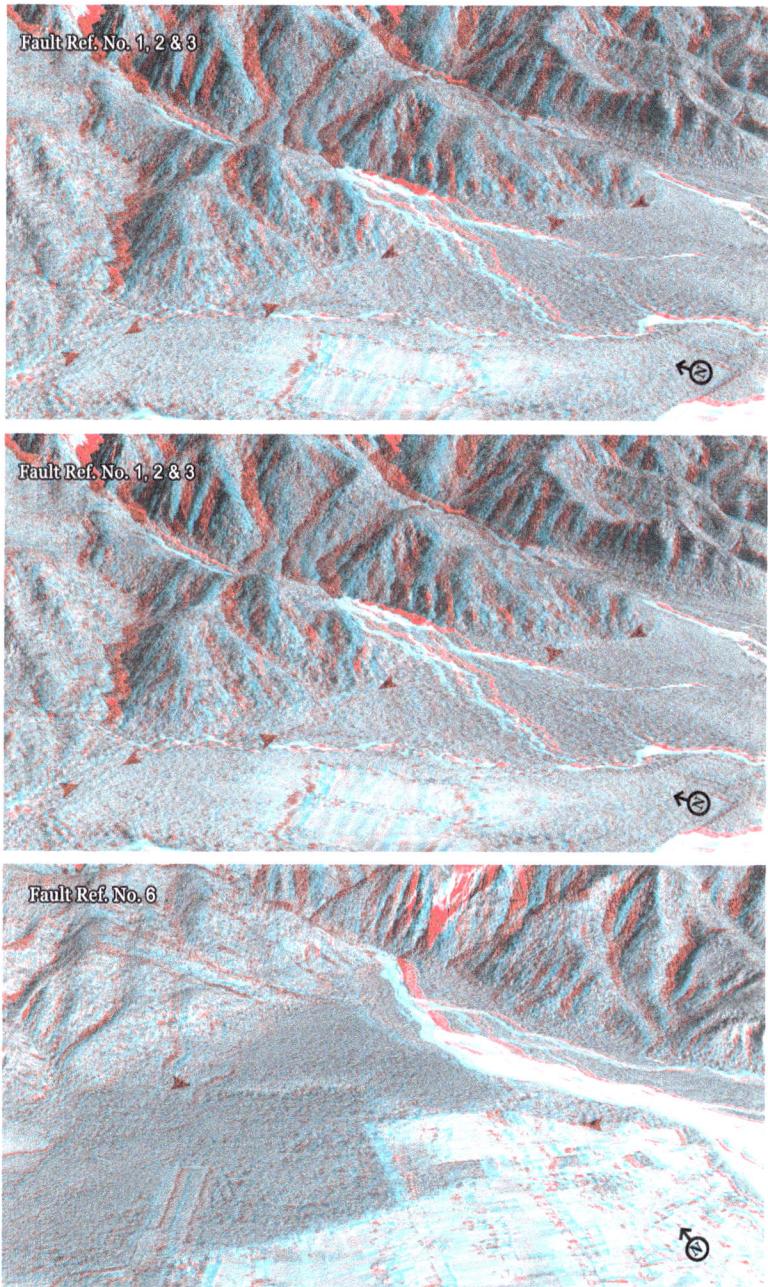

3.5.3 Window-III: Area Between the Nandhaur and Pagboora Rivers

The Sub-Himalayan frontal zone in the area between the Nandhaur and Pagboora rivers is incised by several southward-flowing streams that have produced coalescing fans. Two fault scarps are (numbered 9 and 11) categorized in prominent (Fig. 3.60), and the remaining scarps are inferred. Numerals on fault segments in Fig. 3.60 refer to the fault characteristics given in Table 3.12. The fault scarps are characterized by near-vertical slopes defined by closely spaced contour lines (Fig. 3.61). Two levels of river terraces are observed near the piedmont zone (Fig. 3.62). Digital Elevation Model (DEM) prepared from Cartosat-1A imagery shows a linear feature displacing the contemporary fan surface that defines the active fault at the locality west of the Kumia River (Fig. 3.63). North of Majhola locality, a topographic profile normal to the scarp trend is made (location of profile is shown in Fig. 3.63). The profile shows a height of ~ 12 m corresponding to fault offset (Fig. 3.64).

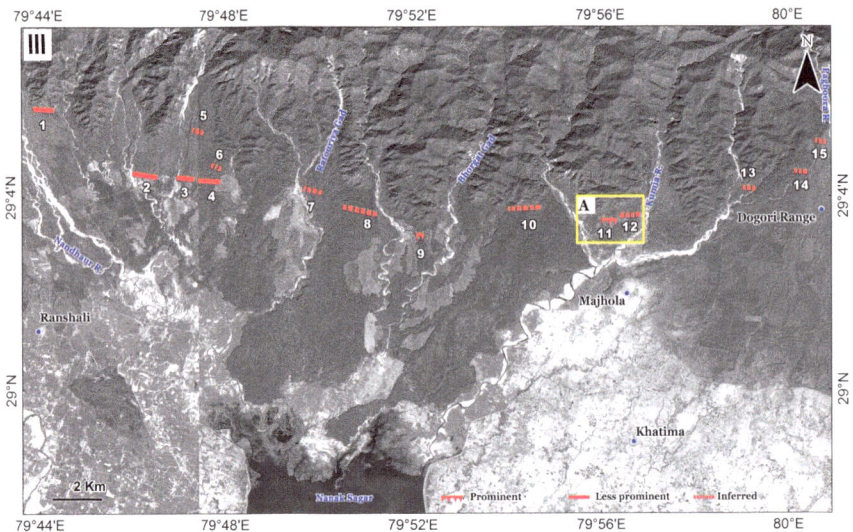

Fig. 3.60 Cartosat-1A grayscale satellite imagery of Window-III showing the active fault segments between the Nandhaur and Pagboora rivers. Contour and geomorphic map of this region is shown in Figs. 3.61 and 3.62. Numerals on active faults (shown in red color) refer to the fault characteristics given in Table 3.12. Rectangle A is a sub-window (Fig. 3.63) location where detailed characteristics of active faults studied

Table 3.12 Characteristics of the fault scarps between Nandhaur and Pagboora rivers

Ref. No	Thrust	Location	Latitude (N)	Longitude (E)	Strike	Length (m)	Fault feature	Fault reference
1	HFT	North of Nandhaur River	29° 05' 45.83"N	79° 44' 05.93"E	NWW–SEE	870	Piedmont fan surface	Uphill or south-facing scarp
2	HFT	Northeast of Nandhaur River	29° 4' 30.47"N	79° 46' 15.88"E	NWW–SEE	924	Piedmont fan surface	Uphill or south-facing scarp
3	HFT	Northeast of Nandhaur River	29° 4' 27.43"N	79° 46' 44.65"E	NWW–SEE	657	Piedmont fan surface	Uphill or south-facing scarp
4	HFT	Northeast of Nandhaur River	29° 4' 25.29"N	79° 47' 37.44"E	E–W	970	Piedmont fan surface	Uphill or south-facing scarp
5	HFT	Northeast of Nandhaur River	29° 04' 52.14"N	79° 47' 18.73"E	NW–SE	312	Terrace	Uphill or south-facing scarp
6	HFT	Northeast of Nandhaur River	29° 04' 38.54"N	79° 47' 55.00"E	NW–SE	286	Terrace	Uphill or south-facing scarp
7	HFT	South of Ratouriya Gad	29° 04' 18.09"N	79° 49' 57.14"E	NW–SE	555	Medial part of fan (old)	Uphill or south-facing scarp
8	HFT	Southeast of Ratouriya Gad	29° 3' 24.61"N	79° 50' 43.10"E	NW–SE	1026	Piedmont fan surface	Uphill or south-facing scarp
9	HFT	Southwest of Bhorgat Gad	29° 03' 19.84"N	79° 52' 08.80"E	E–W	186	Medial part of fan (old)	Uphill or south-facing scarp
10	HFT	Southeast of Bhorgat Gad	29° 3' 40.54"N	79° 54' 20.31"E	NEE–SWW	1014	Piedmont fan surface	Uphill or south-facing scarp
11	HFT	West of Kumia River	29° 03' 37.13"N	79° 55' 56.44"E	E–W	345	Medial part of fan (old)	Uphill or south-facing scarp

(continued)

Table 3.12 (continued)

Ref. No	Thrust	Location	Latitude (N)	Longitude (E)	Strike	Length (m)	Fault feature	Fault reference
12	HFT	West of Kumia River	29° 3' 17.82"N	79° 56' 29.40"E	NEE–SWW	464	Medial part of fan (old)	Uphill or south-facing scarp
13	HFT	Northwest of Dogori Range	29° 04' 12.03"N	79° 59' 11.94"E	NWW–SEE	319	Medial part of fan (young surface)	Uphill or south-facing scarp
14	HFT	Northwest of Dogori Range	29° 04' 30.43"N	80° 00' 19.54"E	NWW–SEE	322	Hill slope/terrace	Uphill or south-facing scarp
15	HFT	West of Pagboora River	29° 04' 58.05"N	80° 00' 39.23"E	NWW–SEE	117	Hill slope/terrace	Uphill or south-facing scarp

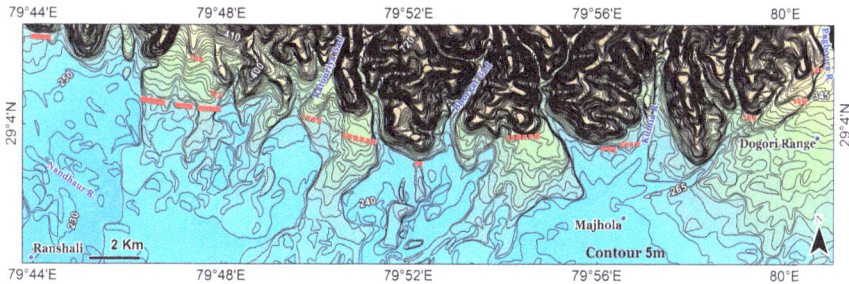

Fig. 3.61 Active fault scarps are aligned along the closely spaced contour lines indicating steepness of the slopes. Contour lines of 5 m interval are derived from Cartosat-1A DEM

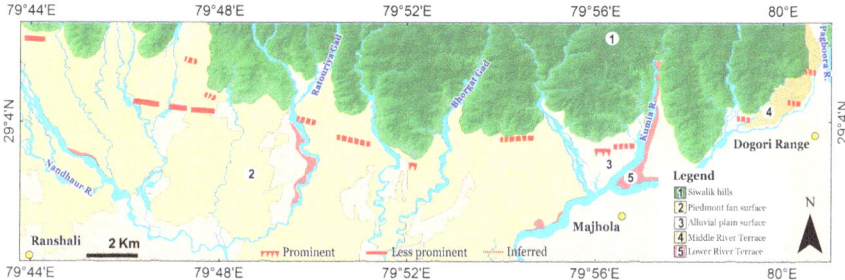

Fig. 3.62 Active fault traces traversing the fan surfaces between the Nandhaur and Pagboora rivers

Fig. 3.63 A three-dimensional perspective view of the fault scarp prepared from Cartosat-1A gray image overlaid on DEM. Red arrows point to the active fault traces in west of Kumia River (fault reference 6 and 7 in Table 3.12). Regional location of the area is shown as rectangle A in Fig. 3.60. Thick line P1 denotes the topographic profile of the scarp constructed across the active fault and is shown in Fig. 3.64

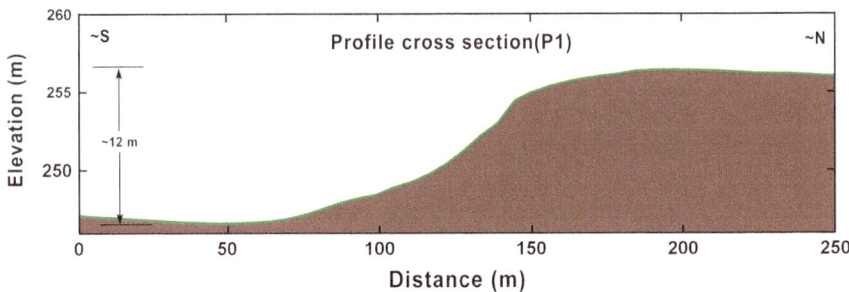

Fig. 3.64 Profile P1 across the fault scarp in northern part of alluvial fan surface near Kumia River gives 12 m vertical separation of the fault. Location is marked in Fig. 3.63

Anaglyph images for active faults between Nandhaur and Pagboora Rivers (Fault Reference number in the figure is corresponds to characteristics of the fault given in Table 3.12).

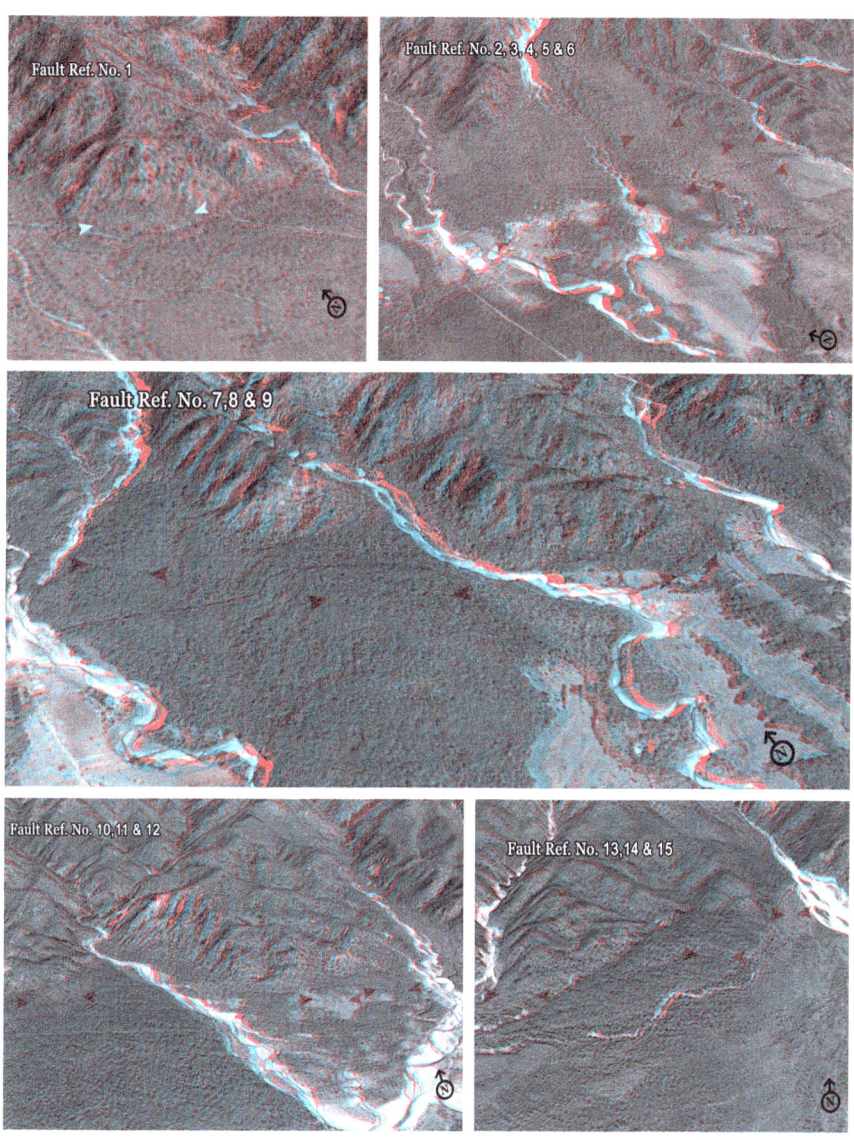

3.6   Tanakpur: Between Pagboora and Kali Rivers

North of Tanakpur in the Sub-Himalaya belt, the Sukhidang–Shiala area, a number of lakelets and paleolakes are formed due to tectonic movement along the Sukhidang–Thuligad Fault (Valdiya 1992). This fault lies between the Dharamsala Formation and the Siwalik Group (Raiverman 2002; Luirei et al. 2014). A series of paleolakes—Talla Banda, Malla Banda, and Shyamal—Tal are described in ascending order from south to north. These paleolake depressions occur on the hanging wall side of north-dipping normal faults forming half-grabens (Luirei et al. 2014). OSL dating of the paleolake sediments yields ages of 278 ± 24 years to 16.3 ± 1.29 ka which implied that the normal faulting took place during late Quaternary indicating reactivation of the MBT. Formation of graben associated with normal faulting occurs in lateral propagation of a thrust fault with left-lateral strike-slip motion and steps over to the right.

Tanakpur lies in the alluvial plain ~ 15 km south of the Sub-Himalayan front, making a concave physiographic bend. South of the Himalayan front, the piedmont zone is made up of coalescing alluvial fans. In the western part of the piedmont zone west of Ammiyan, active fault trace trends WNW–ESE along the Sub-Himalayan front. The fault scarps showing break and steepness in surface slopes are well marked in the Cartosat-1A imagery (Fig. 3.65). Numerals on active faults given in Fig. 3.65 denote the fault characteristic given in Table 3.13. The topographic control of the scarps is well observable in Fig. 3.66. Scarps are aligned along the closely spaced contour intervals indicative of steep almost vertical slopes. In the eastern part, the active fault trending NW–SE is demarcated in the piedmont

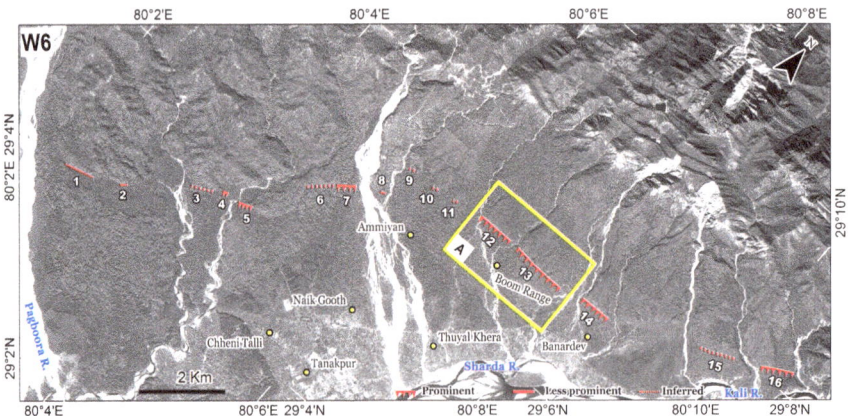

Fig. 3.65 Grayscale Cartosat-1A satellite imagery showing active fault segments between the Pagboora and Kali rivers. Contour and geomorphic map of this region is shown in Figs. 3.66 and 3.67. Numerals on active fault segments (shown as red colored lines) refer to characteristics of the faults given in Table 3.13. Rectangle 'A' sub-window denotes the location where detailed characteristics of active faults studied (Fig. 3.68)

Table 3.13 Characteristics of the fault scarps between the rivers Pagboora and Kali

Ref. No	Thrust	Location	Latitude (N)	Longitude (E)	Strike	Length (m)	Fault feature	Fault reference
1	HFT	West of Chheni Talli	29° 04′ 13.91″N	80° 02′ 28.41″E	NEE–SWW	637	Fault scarp	North of piedmont fan surface
2	HFT	West of Chheni Talli	29° 04′ 26.78″N	80° 02′ 58.51″E	NE–SW	175	Fault scarp	North of piedmont fan surface
3	HFT	Northwest of Chheni Talli	29° 05′ 03.01″N	80° 03′ 45.70″E	NE–SW	477	Fault scarp	North of piedmont fan surface
4	HFT	Northwest of Chheni Talli	29° 05′ 13.41″N	80° 04′ 01.09″E	NE–SW	124	Fault scarp	North of piedmont fan surface
5	HFT	Northwest of Chheni Talli	29° 05′ 17.09″N	80° 04′ 16.11″E	NE–SW	315	Fault scarp	North of piedmont fan surface
6	HFT	Northwest of Naik Gooth/Tanakpur	29° 06′ 05.16″N	80° 04′ 50.22″E	NE–SW	635	Fault scarp	North of piedmont fan surface
7	HFT	Northwest of Naik Gooth/Tanakpur	29° 06′ 17.65″N	80° 05′ 03.83″E	NEE–SWW	217	Fault scarp	North of piedmont fan surface
8	HFT	West of Ammiyan	29° 06′ 30.17″N	80° 05′ 28.38″E	NE–SW	119	Fault scarp	Fan surface and lower river terrace
9	HFT	Northwest of Ammiyan	29° 06′ 59.27″N	80° 05′ 32.50″E	NEE–SWW	175	Fault scarp	Medial part of piedmont fan surface
10	HFT	Northwest of Ammiyan	29° 06′ 59.88″N	80° 05′ 54.59″E	NEE–SWW	176	Fault scarp	Medial part of piedmont fan surface
11	HFT	North of Ammiyan	29° 07′ 02.57″N	80° 06′ 12.02″E	NEE–SWW	155	Fault scarp	Medial part of piedmont fan surface

(continued)

Table 3.13 (continued)

Ref. No	Thrust	Location	Latitude (N)	Longitude (E)	Strike	Length (m)	Fault feature	Fault reference
12	HFT	Northwest of Boom Range	29° 07′ 07.74″N	80° 06′ 47.29″E	E–W	757	Fault scarp	Medial part of piedmont fan surface
13	HFT	Northeast of Boom Range	29° 07′ 05.89″N	80° 07′ 30.27″E	E–W	1169	Fault scarp	Medial part of piedmont fan surface
14	HFT	North of Banardev	29° 07′ 13.45″N	80° 08′ 21.29″E	NEE–SWW	650	Fault scarp	Medial part of piedmont fan surface
15	HFT	Northeast of Banardev	29° 07′ 50.59″N	80° 09′ 49.67″E	NE–SW	784	Fault scarp	Medial part of piedmont fan surface
16	HFT	Northeast of Banardev	29° 08′ 10.23″N	80° 10′ 30.34″E	NE–SW	708	Fault scarp	Medial part of piedmont fan surface

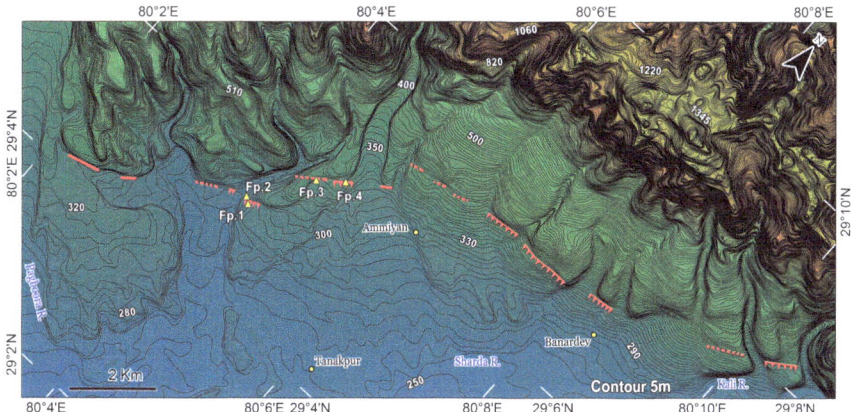

Fig. 3.66 Contour lines clearly demarcate displacement of fan surface between Pagboora and Kali rivers. Contour lines with 5 m interval are derived from Cartosat-1A Digital Elevation Model (DEM) data. Fp-1 to 4 shows location of field photos are provided in Fig. 3.70

Fig. 3.67 Active fault traces of the Himalayan Frontal Thrust offsetting the piedmont fan surfaces between the rivers Pagboora and Kali

zone (Fig. 3.67). A 3D perspective image near the Boom Range shows prominent fault scarps (Fig. 3.68). It represents the sub-window 'A' given in Fig. 3.65 showing location of topographic profile 'P1'. The topographic profiles constructed normal to the fault scarps denote a height or vertical offset of ∼62 m (Fig. 3.69).

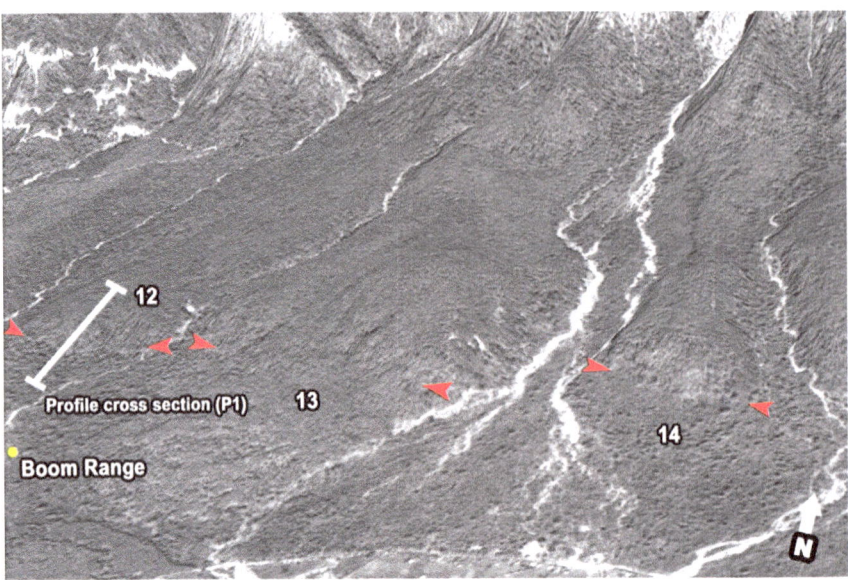

Fig. 3.68 Enlarged view of active fault traces (red lines) around the Boom Range; regional location shown as rectangle 'A' in Fig. 3.65. A 3D perspective image was generated from Cartosat-1A grayscale satellite imagery overlaid on Digital Elevation Model (DEM). Red arrows point to the active fault scarp. Thick black line 'P1' represents the location of topographic profile constructed across the active fault scarp as indicated in Fig. 3.69. DEM image derived from Cartosat-1A stereopair data

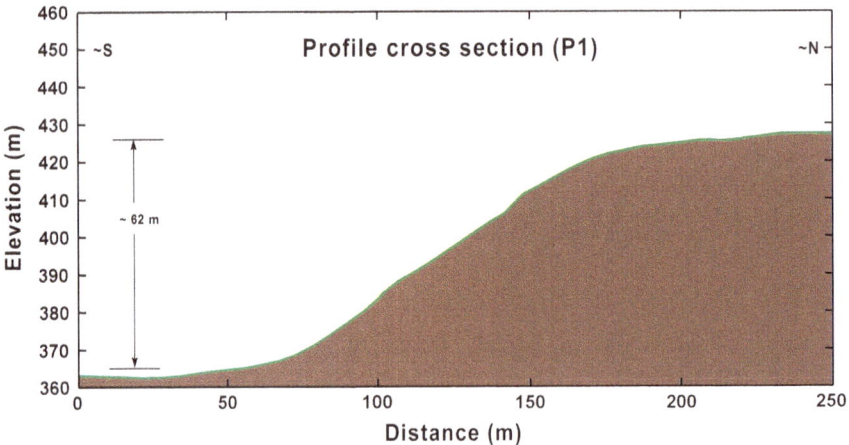

Fig. 3.69 Elevation profile 'P1' normal to the fault scarps indicates ∼62 m vertical separation of the fault. Location is shown in Fig. 3.68

Fig. 3.70 a–d Field photograph showing mapped fault scarp at Fp.1, Fp.2, Fp.3, and Fp.4. Location of these field photographs is shown in Fig. 3.66 (yellow triangle). The field photograph Fp.2 is back side of Fp.1 fault scarp. Bottom photograph Fp.3 shows the eastern extension of fault scarp near the main road between Tanakpur and Pithoragarh. Field photograph Fp.4 showing fault scarp passes through the main road between Tanakpur and Pithoragarh

Anaglyph images for active faults between rivers Pagboora and Kali (Fault Reference number in the figure is corresponds to characteristics of the fault given in Table 3.13).

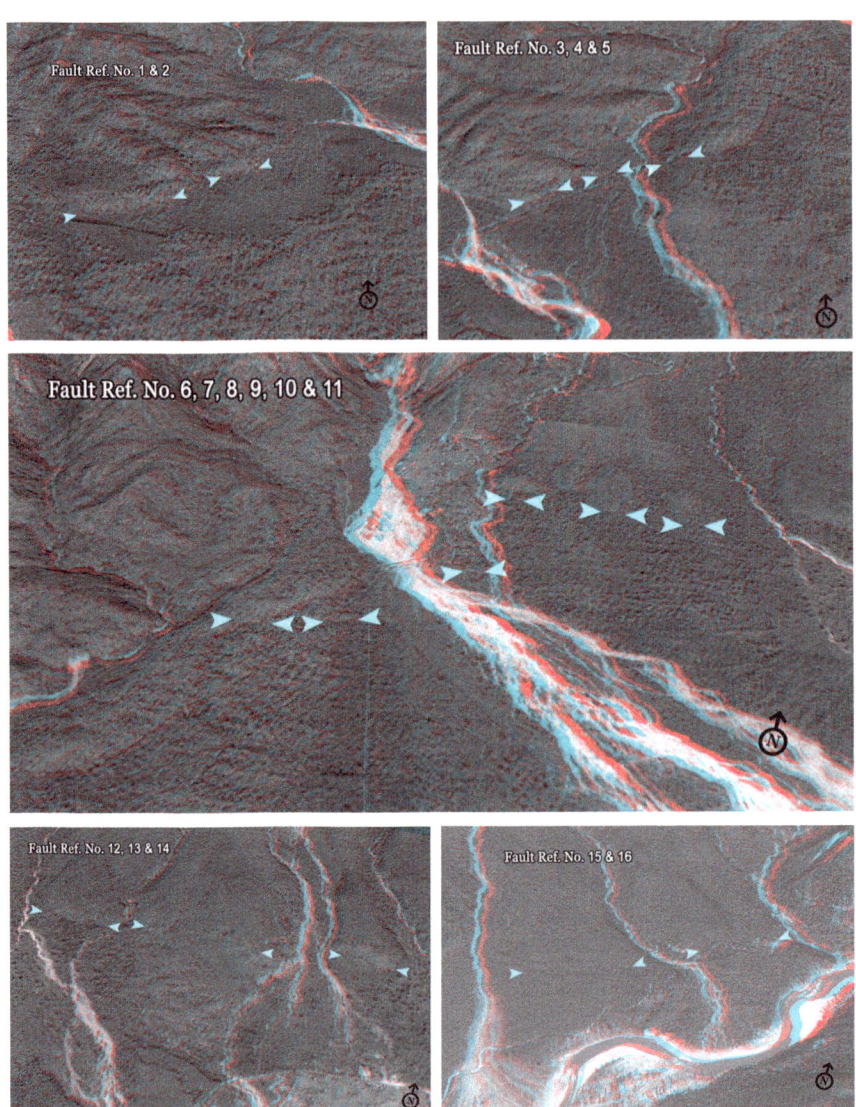

References

Bronk Ramsey C (2009a) Bayesian analysis of radiocarbon dates. Radiocarbon 51(1):337–360

Bronk Ramsey C (2009b) Dealing with outliers and offsets in radiocarbon dating. Radiocarbon 51 (3):1023–1045

Goswami PK, Pant CC (2007) Geomorphology and tectonics of Kota–Pawalgarh Duns, Central Kumaun Sub-Himalaya. Curr Sci 92(5):685–690

Jayangondaperumal R, Daniels Robyn L, Niemi Tina M (2017) A paleoseismic age model for large-magnitude earthquakes on fault segments of the Himalayan Frontal Thrust in the central seismic gap of northern India. Quatern Int. https://doi.org/10.1016/j.quaint.2017.04.008

Kothyari GC, Pant PD, Joshi M, Luirei K, Malik JN (2010) Active faulting and deformation of quaternary landform sub-Himalaya, India. Geochrnometria 37:63–71

Kumar S, Wesnousky SG, Rockwall TK, Briggs RW, Thakur VC, Jayangondaperumal R (2006) Paleoseismic evidence of great surface rupture earthquake along the Indian Himalaya. J Geophys Res 111:B03304. https://doi.org/10.1029/2004JB00309

Luirei K, Bhakuni SS, Srivastava P, Suresh N (2014) Late Pleistocene-Holocene tectonic activities in the frontal part of NE Himalaya between Siang and Dibang river valleys, Arunachal Pradesh, India. Z Geomorp 56:477–493

Mugnier JL, Huyghe P, Gajurel AP, Becel D (2005) Frontal and piggy-back seismic ruptures in the external thrust belt of Western Nepal. J Asian Earth Sci 25:707–717

Nakata T (1972) Geomorphic history and crustal movements of the foothills of the Himalayas. Sci Rep Tohoku Univ 22(7):39–177

Pant MR (2002) A step toward a historical seismicity of Nepal. Adarsa 2:29–60

Philip G, Suresh N, Jayangondaperumal R (2017) Late Pleistocene-Holocene strain release by normal faulting along the Main Boundary Thrust at Logar in the northwestern Kumaun Sub Himalaya, India. Quat Int. http://dx.doi.org/10.1016/j.quaint.2017.05.022

Power PM, Lillie RJ, Yeats RS (1998) Structure and shortening of the Kangra and Dehradun re-entrants, Sub Himalaya, India. Geol Soc Am Bull 110:1010–1027

Raiverman V (2002) Foreland sedimentation in Himalayan tectonic regime: a re-look at the orogenic process. Bishen Singh Mahendra Pal Singh, Dehradun, p 378

Rao YSN, Rahman AA, Rao DP (1973) Wrench-faulting and its relationship to the structure of the southern margin of the Sub-Himalayan belt around Ramnagar, Uttar Pradesh. J Geol Soc India 14(3):249–256

Thakur VC (2013) Active tectonics of himalayan frontal fault system. Int J Earth Sci 102(7):1791–1810

Thakur VC, Pandey AK, Suresh N (2007) Late quaternary-holocene frontal fault zone of the Garhwal Sub Himalaya, NW India. J Asian Earth Sci 29(2/3):305–319

Thakur VC, Jayangondaperumal R, Malik MA (2010) Redefining Wadia-Medlicott's main boundary fault from Jhelum to Yamuna: an active fault strand of the main boundary thrust in northwest Himalaya. Tectonophysics 489:29–42

Valdiya KS (1992) Active Himalayan frontal fault, main boundary thrust and Ramgarh thrust in southern Kumaun. J Geol Soc India 40:509–528

Valdiya KS (2003) Reactivation of Himalayan frontal fault: implication. Curr Sci 85(7):1031–1040

Wesnousky SG, Kumar S, Mahindra R, Thakur VC (1999) Uplift and convergence along the Himalayan Frontal Thrust of India. Tectonics 18:967–976

Chapter 4
Concluding Comments and Structure of Online Interactive Active Fault Database

The Himalaya originated as a result of continental collision between India and south Eurasia with the development of three principal thrust systems, namely the Main Central Thrust (MCT), the Lesser Himalayan duplex and the Main Boundary Thrust (MBT), and the Himalayan Frontal Thrust (HFT) (Yin 2006; Robinson et al. 2006; Jayangondaperumal et al. 2017b). The active plate boundary shifted progressively to the foreland toward the south, with the HFT fault system constituting the contemporary plate boundary. The active faulting and Quaternary deformation are reported in the Sub-Himalayan zone between the MBT and the HFT (Yeats et al. 1992; Thakur 2013). The instrumentally recorded microseismicity is concentrated in a 30–50 km-wide belt over the topographic front of the Higher Himalaya covering the northern Lesser Himalaya in Kumaun–Garhwal and Nepal in the central sector of the Himalaya (Pandey et al. 1999; Arora et al. 2012). The geodetic (GPS) measurements reveal that the segment between the Sub-Himalayan front and the south of Higher Himalayas locked and is undergoing very little internal deformation (Bettinelli et al. 2006; Ader et al. 2012), whereas the Himalayan convergence is consumed through creep motion under and north of the Higher Himalaya, where the Indian plate descends to a greater depth in a ductile regime. The elastic strain accumulated through ongoing convergence in the locked segment is released periodically by large-to-great earthquakes. The earthquake ruptures propagating from the epicenter may remain blind or break the ground surface depending upon depth of the epicenter, magnitude, and inclination of rupture.

In northwest Himalaya, the 2005 M_w 7.6 Kashmir earthquake produced a surface rupture of ~ 75 km showing vertical offset varying from 2.7 to 7 m (Hussain et al. 2009; Avouac et al. 2006). The surface rupture on the ground was along the mapped Balakot Bagh Fault (Hussain et al. 2009; Kaneda et al. 2008), indicating reactivation of the fault. The 2005 earthquake occurred on multiple fault planes invoking active wedge thrust (Bendick et al. 2007). The secondary surface ruptures were mapped in Indian side by Jayangondaperumal and Thakur (2008), whereas in Central Nepal Himalaya, the 2015 M_w 7.9 Gorkha earthquake did not produce any surface rupture. The fault plane solution indicates a near-horizontal reverse fault

© Springer Nature Singapore Pte Ltd. 2018
R. Jayangondaperumal et al., *Active Tectonics of Kumaun and Garhwal Himalaya*,
Springer Natural Hazards, https://doi.org/10.1007/978-981-10-8243-6_4

suggesting that the 2015 Gorkha event was a plate boundary earthquake that occurred over the Main Himalayan Thrust (MHT). The instrumentally recorded data reveal subsurface rupture area of 135×50 km^2, implying that the rupture did not extend up to the Himalayan front (Bilham 2015; Avouac et al. 2015). Evidence of surface ruptures of large and major earthquakes is reported from the Himalayan Frontal Thrust (also called as Main Frontal Thrust) in the western, central, and eastern Himalaya. In Nepal, paleoseismological studies indicate surface ruptures of 1934 and 1255 earthquakes on the MFT, whose epicenters were located in the hinterland of Lesser Himalaya (Sapkota et al. 2013; Bollinger et al. 2014). The 1934 Bihar–Nepal earthquake of magnitude M_w 8.1 inflicted a major damage extending from its epicenter in the middle of Lesser Himalaya around Kathmandu to the Sub-Himalayan front and northern Bihar Gangetic Plain (Dunn et al. 1939; Bilham 1995). The Gorkha earthquake damage was consistent with the area of rupture zone and aftershocks.

Uttarakhand, including Kumaun and Garhwal, lies to the west and adjacent to Nepal. The physiographic and tectonic frameworks of Uttarakhand are analogous to that of Nepal, the former representing western extension of the latter. It is therefore a fair assumption to consider the similar scenario for earthquake hazard assessment in Uttarakhand and Nepal Himalaya. There are accounts including historical large-to-great earthquakes in A.D. 1255, 1344, 1505, 1833, 1934, and 2015 in Nepal (Ambraseys and Douglas 2004; Bollinger et al. 2014; Mugnier et al. 2013; Mishra et al. 2016; Jayangondaperumal et al. 2017a, b). Unlike Nepal, the historical accounts of past earthquakes are lacking in Uttarakhand. The 1803 earthquake affected larger parts of the Alaknanda and Bhagirathi Valleys in the Garhwal Lesser Himalaya. The earthquake damage was also reported from Delhi and Mathura and from an archeological site in the piedmont zone (Thakur 2010). The 1803 earthquake was assigned a magnitude of M_w 7.4 (Ambraseys and Douglas 2004), and later revised to $M_w > 7.5$ and nearly 8.0 (Rajendran et al. 2013, 2015). The paleoseismic evidence for surface rupture of 1803 event is reported on the HFT (Malik et al. 2016). The other source of knowledge about past earthquakes is paleoseismological and archeoseismological investigations. The evidence of an earthquake surface rupture dated at A.D. 1450 is reported in two trenches on the HFT in Garhwal Himalaya (Kumar et al. 2006), and which was correlated as a westward extension of 1505 great earthquake that occurred on the Nepal–Tibet border (Ambraseys and Jackson 2003). A new trench investigation adjoining the earlier Ramnagar trench advocates the 1344 event (Rajendran et al. 2015; Jayangondaperumal et al. 2017b), contradicting the earlier interpretation of 1505 earthquake assigned by Kumar et al. (2006). The 1344 event in Uttarakhand is further corroborated on the basis of archeological monuments (Rajendran et al. 2013) and statistical re-calibration of radiocarbon ages by OxCal (Jayangondaperumal et al. 2017b) and surface faulting along back thrust in the northern limb of Januari anticline. In eastern Nepal, the recurrence interval of great earthquakes($M_w > 8$) is estimated as ~ 750 years, inference based on the surface ruptures of two great earthquakes with $M_w > 8$, 1934 and 1255 reported from the Sub-Himalayan front (Sapkota et al. 2013; Bollinger et al. 2014). In Uttarakhand, for the western section of the Central Seismic Gap in India, Jayangondaperumal et al. (2017b) re-calibrated radiocarbon age model suggests rupture in the earthquake of 1344 CE, with no subsequent large-scale rupture in the last 672 years, reinforcing the concern for an impending large-magnitude event which would be catastrophic for nearby populations.

The Uttarakhand State covers an area of $\sim 53,483$ Km^2 and has a total population of 10.86 million. In this Himalayan State, there are several major rivers with a huge potential for renewable hydroelectric power. Several existing hydropower stations provide electricity to the state and the national grid. There remains a vast potential of more hydropower generation in the Himalayan Rivers. The identification and mapping of active faults have a direct relevance to the location of new power projects, so that the power infrastructure is not located on and close to the active faults. There are pockets of industrial development and population centers in the Tarai region which is close to the Himalayan front. The active nature of the HFT along the front poses earthquake risk to such industrial hubs that needs vulnerability factor to be considered in all construction activity. The 2005 Muzaffarabad earthquake (M_w 7.6) killed about 80,000 people and inflicted financial loss of about 1 billion dollars. The Lesser Himalayan region of Uttarakhand has similar physiographic setting and nearly similar population density. Active fault mapping and earthquake vulnerability analysis can be a mitigation factor in planning and execution of all state-level development plans. The 1999 Chi-Chi earthquake in Taiwan is one of the best studied earthquakes as the area was well instrumented. The earthquake produced extensive surface ruptures of ~ 80 km along the Chelungpo Fault with vertical displacement of 1–8 m. The Chelungpo Fault is imaged in seismic reflection data as the fault dipping 20–30° toward east. The hypocentral and focal mechanism determined is consistent with slips on the Chelungpo Fault. The dense network of strong ground motion data indicates maximum peak ground acceleration (PGA) attaining 1 g and vertical offset of ~ 8 m (Kao and Chen 2000). Extensive damage was observed along this active fault, including collapse of a school building, a bridge over the river and 2000 people getting killed in the Neftegorsk town with a population of 3000. The Chi-Chi earthquake scenario is a good example to construct futuristic earthquake hazard model for an earthquake striking along the Himalayan Frontal Fault system in the frontal Sub-Himalayan zone, where several large infrastructure projects may be located.

As a consequence of active fault mapping carried out in Uttarakhand and the data described in the book, the Wadia Institute of Himalayan Geology, Dehradun, has designed a Quaternary fault and fold database for frontal Garhwal and Kumaun Himalaya in conformance with standards defined by active fault database for the National Quaternary Fault and Fold Database. Himalaya is the most seismically active region of India; however, little information exists on the location, style of deformation, and slip rates of Quaternary faults. Thus, to provide an accurate, user-friendly, reference-based fault inventory to the public, a GIS shapefile of Quaternary fault traces and compiled attributes describe age, relative activity, and general parameters for each fault or fold. This book describes relevant information pertaining to the shapefile and online access and availability. The database is the first comprehensive digital compilation of Quaternary faults in the Himalaya, and future updates will include the results of new research. This is novel attempt to provide information to the society regarding the location of active faults which might be hazardous for construction of buildings and other infrastructures.

4.1 Web-Based Himalayan Active Fault Database Interface System (WHADIS)

The Web-based Active Fault Database Interface System (WHADIS) is a geodata repository of Quaternary active faults of the Garhwal and Kumaun Himalaya. The online database is structured, new data has been added and complied by the authors and is devoted to the scientific community for the seismic assessment at a regional and national scales. It is an open source interactive database, and it can be utilized without restriction. The database server is maintained by the Wadia Institute of Himalayan Geology (WIHG), Dehradun, with the technical support received from Amigo Optima Pvt. Ltd. This online interactive database is customized in such a way that users can understand the seismogenic faulting in the Himalayan foothills. The database can be accessed through http://www.wihg.res.in/whadis-himalaya from commonly available Web browsers such as Mozilla Firefox, Internet Explorer and Google Chrome. The database contains information about its location, geo-coordinates, strike, age, length, fault type, and reference.

4.1.1 Online Database Structure

The online database has followed successive phases that have led to access the contents present in the Web server. The first phase comprises the base layer which shows the physiographic structure of the earth surface. The base layer is open source natural earth terrain data in 10-m scale (the map includes shaded relief, water, and drainages). The second phase comprises three parts: The first part covers list of windows that are explained in detail in the previous chapters; the second part covers age classification of active faults. The line color demarcates the fault age based on the Quaternary surface deformation; the third part provides details about overlay layers which include fault traces, Cartosat–1A gray imagery, high-resolution contour lines, and geomorphic surface features. The on-screen visualization can be enhanced from the zoom tool presented in the left corner or by scrolling the mouse. The design and specification of the database are described in Table 4.1. The default appearance of the Web interface is presented in Fig. 4.1. The default view shows physical terrain surface as base layer along with overlaid fault traces of Window–1. Users can add overlay features by using the check or un-check radio buttons based on their convenience.

Table 4.1 Specifications of the online interactive database system (WHADIS)

S. no.	Dataset	Display features	Data source
1.	Base layer	1. Terrain surface (default) 2. Cartosat–1A image	1. Types of base layer: Global map, Google map, Google satellite, Google hybrid, Google physical and Google streets 2. Cartosat–1A is purchased from NRSA, India (http://uops.nrsc.gov.in)
2.	Fault trace	1. Thrust name 2. Location 3. Geocoordinates 4. Strike direction 5. Fault age 6. Length in meters 7. Fault type and reference	Fault lines are mapped based on the existing literatures, trench excavation results, and geomorphic surface displacement
3.	Geomorphic map*	1. Hilly terrain 2. Different types of surfaces 3. Level of river terraces 4. Alluvial and piedmont fan surfaces 5. Present river bed	Geomorphic features are mapped based on the Cartosat–1A 3D visual interpretation, existing map reference and closeness of contour lines
4.	Contour lines*	Contour lines with 5 or 10 m interval	Contours are derived from Cartosat-based Digital Elevation Model (DEM) in GIS environment
5.	Prominent fault details*	1. Fault profile cross section 2. Field photographs	1. Fault profile derived from DEM 2. Photographs captured at the time of field visit

* Data will be available to the readers on requisition basis

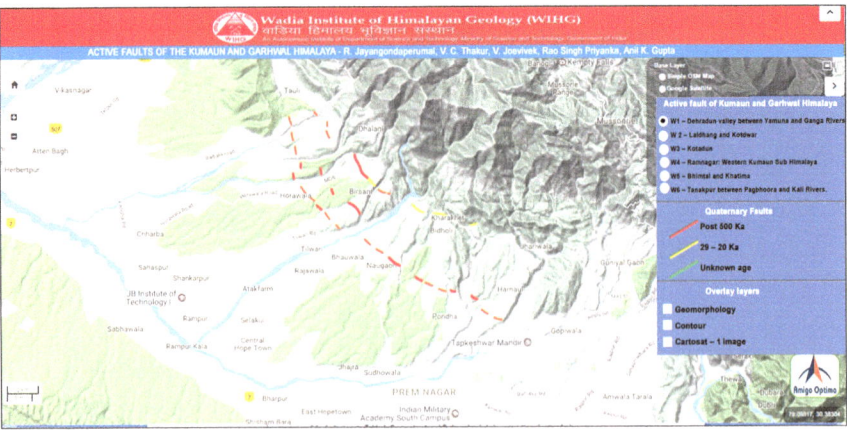

Fig. 4.1 Default appearance of the active fault database in the Web browser

4.2 Potential Usage

The purpose of developing open source fault database is to fulfill the needs of a broad group of users, ranging from the scientific to general community. It helps to understand the spatial and temporal relationships of active faults at local and regional level. It provides geographic visualization for modeling surface deformation and plate strain distribution in large scale. This will act as a guide for paleoseismologists and earthquake geologists to identify areas that are needed for trenching studies. Similarly, the local and state planners can utilize this database to infer the earthquake-prone zones that helps to sustain critical infrastructure near active faults. This database will act as a user-friendly online interface and create awareness to the people to avoid man-made activities around the active fault zone.

The present fault database system allows geological community to update new faults which is observed from the Web interactive platform. They can submit inferred fault location along with detailed characteristics of the fault to the authors of this book. All submitted data will be subjected to review; after validation, it will be updated in the WHADIS database server.

Acknowledgements Database Access: The active faults displayed on this interactive Web browser cannot be used for commercial purpose. Contact: Wadia Institute of Himalayan Geology (WIHG) or authors of this book when you wish to reprint or reproduce in any manner. Permission to use this data is granted to full acknowledgment of the following source.

R. Jayangondaperumal, V. C. Thakur, V. Joevivek, Rao Singh Priyanka, Anil K. Gupta (2018), Active faults of the Kumaun and Garhwal Himalaya, Springer Natural Hazards series, Springer publishers, 147 p. doi:10.1007/978-981-10-8243-6.

References

Ader T, Avouac JP, Liu-Zeng J, Lyon-Caen H et al (2012) Convergence rate across the Nepal Himalaya and interseismic coupling on the Main Himalayan Thrust: implications for seismic hazard. J Geophys Res 117:403. https://doi.org/10.1029/2011jb009071

Ambraseys N, Jackson D (2003) A note on early earthquakes in northern India and southern Tibet. Curr Sci 84(4): 570–582

Ambraseys N, Douglas J (2004) Magnitude calibration of North Indian earthquakes. J Geophys Int 159:165–206

Arora BR, Gahalaut VK, Kumar N (2012) Structural control on a seismicity of northwest Himalaya. J Asian Earth Sci 57:15–24

Avouac J-P, Ayoub F, Lefrince S, Konca K, Hellembergner DV (2006) The 2005, Mw 7.6, Kashmir earthquake: sub-pixel correlation of ASTER images and seismic wave form analysis. Earth Planet Sci Lett 249:514–528

Avouac J-P, Meng L, Wei S, Wang T, Ampuero J-P (2015) Lower edge of locked Main Himalayan Thrust unzipped by the 2015 Gorkha earthquake. Nat Geosci 8:708–711

Bendick R, Bilham R, Khan MA, Khan SF (2007) Slip on an active wedge thrust from geodetic observations of the 8 October 2005 Kashmir earthquake. Geology 35(3):267–270

Bettinelli P, Avouac JP, Flouzat M, Jouanne F, Bollinger L, Willis P, Chitrakarm G (2006) Plate motion of India and Interseismic strain in the Nepal Himalaya from GPS and DORIS measurements. J Geod 80:567–589

Bilham R (1995) Location and magnitude of the 1833 Nepal earthquake and its relation to the rupture zones of contiguous great Himalayan earthquakes. Curr Sci 69:101–128

Bilham R (2015) Raising Kathmandu, Nat Geosci 8:582–584

Bollinger L, Sapkota SN, Tapponnier P, Klinger Y, Rizza M, Van der Woerd J, Tiwari DR, Pandey R, Bitri A, Bes de Berc S (2014) Estimating the return times of great Himalayan earthquakes in eastern Nepal: evidence from the Patu and Bardibas strand of the Main Frontal Thrust. J Geophys Res Solid Earth 119:7123–7163. https://doi.org/10.1002/2014jb01090

Dunn JA, Auden JB, Ghosh AMN, Roy SC (1939) The Bihar-Nepal earthquake of 1934. Mem Geol Survey India 73:391 (reprinted 1981)

Hussain A, Yeats RS, MonaLisa (2009) Geological setting of the 8 October 2005 Kashmir earthquake. J Seism 13(3):315–325

Jayangondaperumal R, Daniels Robyn L, Niemi Tina M (2017b) A paleoseismic age model for large-magnitude earthquakes on fault segments of the Himalayan Frontal Thrust in the Central Seismic Gap of northern India. Quatern Int. https://doi.org/10.1016/j.quaint.2017.04.008

Jayangondaperumal R, Kumahara Y, Thakur VC, Kumar A, Srivastava P, Shubhanshu D, Joevivek V, Dubey AK (2017a) Great earthquake surface ruptures along backthrust of the Janauri anticline, NW Himalaya. J Asian Earth Sci 133:89–101. https://doi.org/10.1016/j.jseaes.2016.05.006

Jayangondaperumal R, Thakur, VC (2008) Co-seismic secondary surface fractures on southeastward extension of the rupture zone of the 2005 Kashmir earthquake. Tectonophysics 446:61–76

Kaneda H, Nakata T, Tsutsumi H, Kondo S, Sugito N, Awata Y, Akhtar S, Majid A, Khatak W, Awan A, Yeats RS, Hussain A, Ashraj M, Wesnousky SG, Kausar B (2008) Surface rupture of the 2005 Kashmir, Pakistan earthquake and its active tectonic implication. Bull Seism Soc Am 98:512–557

Kao H, Chen WP (2000) The Chi-Chi earthquake sequence: active, out-of-sequence thrust faulting in Taiwan. Science 288:2346–2349

Kumar S, Wesnousky SG, Rockwall TK, Briggs RW, Thakur VC, Jayangondaperumal R (2006) Paleoseismic evidence of great surface rupture earthquake along the Indian Himalaya. J Geophys Res 111:B03304. https://doi.org/10.1029/2004JB00309

Malik JN, Naik SP, Sahoo S, Okumura K, Mohanty A (2016) Paleoseismic evidence of the CE 1505 (?) and CE 1803 earthquakes from the foothill zone of the Kumaon Himalaya along the

Himalayan Frontal Thrust (HFT), India. Tectonophysics. http://dx.doi.org/10.1016/j.tecto. 2016.07.026

Mishra RL, Singh I, Pandey A, Rao PS, Sahoo HK, Jayangondaperumal R (2016) Paleoseismic evidence of a giant medieval earthquake in the eastern Himalaya. Geophys Res Lett 43:5707–5715. https://doi.org/10.1002/2016GL068739

Mugnier JL, Gajure A, Huyghe P, Jayangndaperumal R, Jouanne F, Upreti BN (2013) Structural interpretation of the great earthquakes of the last millennium in the central Himalaya. Earth Sci Rev 127:30–47

Pandey MR, Tandulkar RP, Avouac JP, Vergne J, Heritier T (1999) Seismotectonics of the Nepal Himalaya from a local seismic network. J Asian Earth Sci 17:703–712

Rajendran CP, John B, Rajendran K (2015) Medieval pulse of great earthquakes in the central Himalaya: viewing past activities on the frontal thrust. J Geophys Res Solid Earth 120. https://doi.org/10.1002/2014jb011015

Rajendran CP, Rajendran K, Sanwal J, Sandiford M (2013) Archeological and historical database on the medieval earthquakes of the central Himalaya: ambiguities and inferences. Seism Res Lett 84(6):1–11. https://doi.org/10.1785/0229130077

Robinson DM, DeCelles PG, Copeland P (2006) Tectonic evolution of the Himalayan thrust belt in western Nepal. Geol Soc Am Bull 118:865–885. https://doi.org/10.1130/b25911.1

Sapkota SN, Bollinger L, Klinger L, Tapponnier P, Gaudemer Y, Tewari D (2013) Primary surface ruptures of the great Himalayan earthquakes in 1934 and 1255. Nat Geosci 6:71–76

Thakur VC (2010) Geo-archeology at Khajnawar in western Uttar Pradesh plain. Curr Sci 98(8): 1112–1119

Thakur VC (2013) Active tectonics of Himalayan Frontal Fault system. Int J Earth Sci 102(7): 1791–1810

Yeats RS, Nakata T, Farah A, Mirza MA, Pandey MR, Stein RS (1992) The Himalayan Frontal Fault system. In: Bucknam RC, Hancock PL (eds) Major active faults of the world: results of IGCP Project 206: annales tectonicae, supplement 5, 6, pp 85–98

Yin A (2006) Cenozoic tectonic evolution of the Himalayan orogen as constrained by along–strike variation of structural geometry, exhumation history, and foreland sedimentation. Earth Sci Rev 28:211–280

Printed by Printforce, the Netherlands